Analysis of Antibiotic/Drug Residues in Food Products of Animal Origin

Analysis of Antibiotic/Drug Residues in Food Products of Animal Origin

Edited by

Vipin K. Agarwal

The Connecticut Agricultural Experiment Station
New Haven, Connecticut

Springer Science+Business Media, LLC

Library of Congress Cataloging-in-Publication Data

Analysis of antibiotic/drug residues in food products of animal origin
    edited by Vipin K. Agarwal.
       p.   cm.
    "Proceedings of an American Chemical Society Agricultural and food
  Chemistry Division Symposium on Antibiotic/Drug Residues in Food
  Products of Animal Origin, held August 25-30, 1991, in New York, New
  York"--T.p. verso.
    Includes bibliographical references and index.
    ISBN 978-1-4613-6480-1   ISBN 978-1-4615-3356-6 (eBook)
    DOI 10.1007/978-1-4615-3356-6
    1. Veterinary drug residues--Analysis--Congresses. 2. Antibiotic
  residues--Analysis--Congresses.   I. Agarwal, Vipin K., 1955-   .
  II. American Chemical Society. Division of Agricultural and Food
  Chemistry.   III. American Chemical Society Agricultural and Food
  Chemistry Division Symposium on Antibiotic/Drug Residues in Food
  Products of Animal Origin (1991 : New York, N.Y.) IV. Title:
  Analysis of anitbiotic drug residues in food products of animal
  origin.
    [DNLM: 1. Antibiotics--analysis--congresses. 2. Drug Residues-
  -analysis--congresses. 3. Food Contamination--congresses. 4. Meat-
  -analysis--congresses.   WA 701 A352 1991]
  RA1270.V47A53   1992
  664'.07--dc20
  DNLM/DLC
  for Library of Congress   92-12153
                                                                  CIP

Proceedings of an American Chemical Society Agricultural and Food
Chemistry Division Symposium on Antibiotic/Drug Residues in Food
Products of Animal Origin, held August 25-30, 1991, in New York, New York

ISBN 978-1-4613-6480-1

© 1992 Springer Science+Business Media New York
Originally published by Plenum Press New York in 1992
Softcover reprint of the hardcover 1st edition

PREFACE

In the last three decades, use of antibiotics/drugs in animal husbandry programs has grown tremendously. Antibiotics/drugs are used therapeutically to cure diseases, and subtherapeutically to control the outbreak of diseases, improve feed efficiency and promote growth.

The presence of antibiotic/drug residues in food products of animal origin, i.e., meat, poultry and milk, can be a potential health hazard to consumers. Significant research is being done to develop new methods or to improve on existing methods to confirm and quantitatively determine the antibiotic/drug residues in meat, poultry and milk.

This book covers recent development and application of various analytical techniques for the determination of antibiotic/drug residues in food products of animal origin.

I thank the authors for their time and efforts in preparing the manuscripts and all the reviewers for reviewing the manuscripts. I also thank to the Division of Agricultural and Food Chemistry of the American Chemical Society for sponsoring the symposium and Hewlett Packard, Palo Alto, CA, Perkin Elmer Corp., Norwalk, CT, Millipore Corporation, Milford, MA, and Kraft General Foods, Glenview, IL, for their generous financial support for the symposium.

Vipin K. Agarwal
New Haven, CT

CONTENTS

Importance of Laboratory Validations and
      Accurate Descriptions of Analytical
      Procedures for Drug Residues in Foods.................. 1
  C. J. Barnes

Incidence of Residues in Foods of Animal Origin............. 5
  M. S. Brady and S. E. Katz

Radioimmunoassay of Antibiotic/Drug
2   Residues in Foods of Animal Origin...................23
  S. J. Steiner and J. N. Harris

Current Problems Associated with the
      Detection of Antibiotic/Drug Residues
      in Milk and Other Food................................31
  S. E. Charm

ELISA and Its Application for Residue
      Analysis of Antibiotics and Drugs in
      Products of Animal Origin.............................57
  D. E. Dixon-Holland

Evaluation and Testing of Charm Test II
      Receptor Assays for the Detection of
      Antimicrobial Residues in Meat.......................75
  G. O. Korsrud, C. D. C. Salisbury,
  A. C. E. Fesser and J. D. MacNeil

Immunochemical Methods In the Analysis
      of Veterinary Drug Residues...........................81
  N. Haagsma and C. van de Water

Residues of Anabolic Steroids in
      Muscle Tissues from Carcasses with
      Positive Injection Sites..............................99
  C. H. Van Peteghem, E. Daeseleire
  and A. De Guesquiere

Approaches to the Detection and
      Confirmation of Drug Residues in Milk.................107
  M. C. Carson, D. N. Heller,
  P. J. Kijak, and M. H. Thomas

The Application of Matrix Solid Phase
        Dispersion (MSPD) to the Extraction and
        Subsequent Analysis of Drug Residues
        in Animal Tissues......................................119
    S. A. Barker and A. R. Long

Liquid Chromatographic Approaches to
        Determination of B-Lactam Antibiotic
        Residues in Milk and Tissues...........................133
    W. A. Moats

Comparison of Chromatographic Procedures
        for Determining Residues of Penicillins................147
    M. Petz

A High Performance Liquid Chromatographic
        Method for the Determination of Nine
        Sulfonamides in Milk...................................165
    V. K. Agarwal

Sulfadimethoxine and Sulfamonomethoxine
        Residue Studies in Chicken Tissues and Eggs............173
    T. Nagata, M. Saeki, T. Ida and M. Waki

The Current Overview of Feed Additives
        and Veterinary Drugs and Their Residual
        Analysis in Japan......................................187
    H. Nakazawa, M. Fujita and M. Horie

Application of Electrochemical and
        UV/Visible Detection to the LC Separation
        and Determination of Methylene Blue and
        Its Demethylated Metabolites from Milk.................197
    J. E. Roybal, R. K. Munns, D. C. Holland,
    J. A. Hurlbut and A. R. Long

The Presence of Dihydroerythroidines in the
        Milk of Goats Fed Erythrina poeppigiana
        and E. berterona Foliage...............................211
    L. D. Payne and J. P. Foley

Analysis of Selected Chemotherapeutics and
        Antiparasitics...225
    R. Malisch

Contributors...253

Index..259

IMPROTANCE OF LABORATORY VALIDATIONS AND ACCURATE DESCRIPTIONS OF

ANALYTICAL PROCEDURES FOR DRUG RESIDUES IN FOODS

Charlie J. Barnes

Food and Drug Administration
Beltsville, MD 20705

INTRODUCTION

Among drug residue analysts, strong disagreements about the validity and interpretation of analytical data or the superiority of one analytical procedure over another frequently occur. I have frequently observed these differences between persons or groups involved in the interpretation of data from the monitoring of animal food products for drug residues, or the selection of drug residue procedures for interlaboratory evaluation and approval. These disagreements can consume considerable time, result in unfavorable publicity, and cause considerable confusion among persons with good intentions. Often these disagreements result from the (1) lack of data and agreement on evaluation criteria, (2) careless use of language, or (3) the absence of adequate interpretation of analytical data. I think we need to pause periodically and remind ourselves of some of (1) the laboratory data and practices needed to establish the performance characteristics of analytical methods, (2) the importance of taking the time to generate this data, and (3) the need to accurately and effectively communicate the performance characteristics of analytical procedures. These performance characteristics need to be understood by the persons asking for the analyses and those who will make decisions based on the resulting analytical data.

BACKGROUND

Prior to the 1970's, animal drug sponsors sometimes selected the parent drug on which to base residue monitoring methods without giving adequate consideration to metabolites. Now drug residue scientists are likely to identify the major metabolites; generate data on the rate of depletion of parent drug and the metabolites in the tissues of the animal; select the tissue in which the residues deplete the slowest; and select either the parent compound, one of its metabolites, or a fragment common to the parent and metabolites as a marker compound. From the depletion data for the parent and metabolites, the level of this marker compound that corresponds to the safe level of total residues in the animal can be determined. This level of the marker is referred to as Rm.

Another term used in animal drug methodology is "incurred sample." This is used to designate animal tissue or fluid samples that contain the marker compound as a result of animal dosing and physiological depletion.

Analysis of Antibiotic Drug Residues in Food Products of Animal Origin
Edited by V.K. Agarwal, Plenum Press, New York, 1992

1

VALUATION PARAMETERS

Much has been written, or debated about the evaluation of analytical procedures in the last ten years. Examples are reports by Keith[1] et al, Horowitz[2], Cardone[3], the Food and Drug Administration[4], and the Association of Official Analytical Chemists[5]. The principal criteria by which quantitative analytical methods are usually evaluated are (1) accuracy, (2) specificity, (3) precision, and (4) practicability. During the development of analytical methods that you expect to routinely use in making decisions with significant health or economic impact, these four parameters should be thoroughly investigated.

Accuracy

Accuracy is measured by the percentage of the marker compound that can be recovered from fortified control or incurred samples. Fortified controls are easy to produce and the data resulting from their use are helpful in evaluating methods, or your progress towards the development of a method. However, incurred samples are the "real" samples, and should be used in the development of your method. An analytical procedure may not perform the same in fortified samples as it does in incurred samples. There may be problems with interferences from drug residues other than the marker compound. The marker compound may not have the same stability in fortified samples as it does in incurred tissues. Well designed studies using tissue from animals that have been dosed with radiolabeled drugs or exhaustive extraction studies are frequently helpful in demonstrating that a proposed method is quantitatively recovering the marker from incurred samples.

Precision

Precision is a measure of the interlaboratory and intralaboratory variability in repeated measurements. An effective way to convey information about precision is to describe the expected intralaboratory and interlaboratory coefficient of variation of data produced by the analytical method. During method development, the various steps of the method should be tested and improved until they work well enough to accomplish their intended purpose.

Specificity

Specificity is a measure of the ability of an analytical procedure to respond only to the substance of interest. It is frequently difficult to make good estimates of specificity. During method development, it is sometimes practical to examine the physical and chemical properties of the drugs and additives that are likely to be used in the animal species involved, estimate which of these have the potential to interfere, and run these through the determinative and confirmatory procedures. This may at least establish the ability of the determinative and confirmatory procedures to distinguish the marker from the group of compounds studied. However, you should remain aware that interferences in animal tissues may originate from a large number of sources. Analyses of control samples from the target species animals raised under different conditions of animal husbandry are probably helpful in the evaluation of the potential for specific interferences or general background from the sample matrix.

Many analysts recommend that positives or over tolerance drug levels that are found and quantitated in samples of unknown history by determinative procedures be subjected to additional procedures for confirming their identity. Depending upon the certainty with which one needs to confirm the identity of positives one can use a single second analytical procedure, a battery of tests based on different physicochemical principles, or procedures such as mass spectrometry which can provide considerable structural details of the suspect compound. Many analysts in the animal drug residue area are now comfortable with analytical methods that combine a quantitative thin layer chromatographic, liquid chromatographic (LC), or gas chromatographic (GC) determinative procedure with a combined GC/ or LC/mass spectrometry procedure for confirming the identity of positives. Frequently, portions of the same sample extract can be used for both the determinative and confirmatory procedures. Where sufficient levels of drug residues are available, combined GC/infrared spectrometry might be a definitive confirmatory procedure. The application of a battery of tests based on different physicochemical principles can provide substantial evidence that a positive has been correctly identified.

Practicability

Although perhaps of lesser importance, there are several other attributes that may make one method more attractive than another. We refer to this as the practicability of the method. The method should (1) utilize commercially available reagents and instrumentation, (2) be capable of being performed by experienced animal drug residue analysts in a reasonable amount of time, (3) not cost too much per sample to perform, and (4) be capable of being performed safely. It is very important that the stability and safe storage conditions of the sample extracts and standards, and safe stopping points in the methods are detailed.

VALIDATION DATA

After your preliminary studies indicate that you have developed a method that meets your specificity, accuracy, and precision requirements, you should, at least, subject it to a validation series in your laboratory. If problems are encountered, the validation should be stopped, the problems should be corrected by modifications in the method, and the modified method subjected to a complete and new validation series. If at all practical, methods that have been successfully validated in your laboratory should be subjected to an interlaboratory validation series; especially, if they are to be used repeatedly in animal drug residue monitoring. For quantitative methods for drugs in animal tissues, the Center for Veterinary Medicine, Food and Drug Administration recommends the following sample series for validation studies in your laboratory, or for each of participants in an interlaboratory validation study:

        Determinative Procedure Samples
        (1) 5 controls
        (2) 5 controls fortified with marker at 1/2 half Rm
        (3) 5 controls fortified with marker at 1 Rm
        (4) 5 controls fortified with marker at 2 Rm
        (5) 5 incurred samples with marker at approximately 1 Rm

        Confirmatory Procedure Samples - use 1, 3, and 5 above.

It is helpful to think of the steps in establishing a new analytical method as follows: (1) development of the method, (2) validation in your laboratory, (3) validation in an interlaboratory trial, (4) routine utilization, and (4) the implementation a method/analyst performance monitoring program.

CONCLUSIONS

Validate analytical methods, at least in your laboratory, and if possible, in an interlaboratory study before routinely using them to generate critical animal drug residue monitoring data. Along with the samples analyzed each day, generate at least a limited amount of data which show that the method is operating properly. Use accuracy, precision, and specificity to describe the performance characteristics of your quantitative analytical methods, and that of others. Detail the concentration ranges in which your method is suitable for the preliminary detection and also for the quantitation of residues. Provide technical interpretations of the analytical results you obtain.

REFERENCES

1. L. K. Keith, W. Crummet, J. Deegan Jr., R. A. Libby, J. K. Taylor, and G. Wentler, Principles of Environmental Analysis, Anal. Chem. 55: 2210 (1983).
2. W. Horwitz, Evaluation of Analytical Methods Used for Regulations of Foods and Drugs, Anal. Chem. 54:67A (1982).
3. M. J. Cardone, Detection and Determination of Error in Analytical Methodology. Part I. In the Method Verification Program, J. Assoc. Off. Anal. Chem. 66:1257 (1983).
4. Association of Official Analytical Chemists, Guidelines for Collaborative Study Procedure to Validate Characteristics of a Method of Analysis, J. Assoc. Off. Anal. Chem. 72:694 (1989).
5. Food and Drug Administration, Sponsoring Compounds Used in Food Producing Animals; Availability of Guidelines for Human Food Safety Evaluation, Federal Register:4495889 (1987).

INCIDENCE OF RESIDUES IN FOODS OF ANIMAL ORIGIN

Marietta Sue Brady and Stanley E. Katz

Department of Biochemistry and Microbiology
Cook College/New Jersey Agricultural Experiment Station
Rutgers - the State University of New Jersey
New Brunswick, New Jersey 08903-0231

ABSTRACT

Antibiotic/Antimicrobial residues occur in foods of animal origin as an unwanted concomitant of drug use in animal agriculture. Residues occur in all types of food of animal origin including milk, muscle tissue and organ tissue, due to the widespread use of antibiotics/antimicrobials for growth stimulation and disease prevention and treatment. Data on the incidence of residues in animals from the statistically-based monitoring programs, surveillance programs and in-plant testing performed by the FSIS/USDA are presented for 1988 and 1989. The results of spot market surveys of residues in milk for 1989-1990 are presented. Reasons for the frequency of residues and their potential biological significance are discussed.

INTRODUCTION

Antibiotics have been used extensively in animal agriculture for some 35 years. Of the 2.5 million kilograms of antimicrobial agents used annually in the United States, approximately 45% is used in animal agriculture. It has been estimated that 80% of the poultry, 75% of the swine, 60% of feedlot cattle and 75% of calves have been fed at least one antibiotic or antimicrobial at some time during their lifetime [1]. Table 1 shows some of the more common antibiotics and antimicrobials currently used in feeds [2].

The occurrence of antibiotic and antimicrobial residues in products of animal origin is an expected event since residues are a concomitant of the use of drugs for growth stimulation, as well as prevention and treatment of disease in animals. Residues can be defined as trace amounts of a compound that remain in a food product following approved use, and are not considered harmful when ingested as long as the concentration is below a toxicologically approved level [3]. Residues

Analysis of Antibiotic Drug Residues in Food Products of Animal Origin
Edited by V.K. Agarwal, Plenum Press, New York, 1992

5

TABLE 1. Partial list of Commonly Used Antibiotics and Antimicrobials in Animal Feeds.

Drug	Animal Species

Bacitracin	Chickens, turkeys, swine, cattle
Bambermycins	Chickens, turkeys, swine
Chlortetracline*	Chickens, turkeys, swine, cattle, sheep
Erythromycin*	Chickens, turkeys
Hygromycin B	Chickens, swine
Lasalocid	Chickens, swine
Lincomycin	Chickens, swine
Monensin	Chickens, cattle
Neomycin*	Chickens, turkeys swine, sheep, cattle
Nystain	Chickens, turkeys
Oleandomycin	Chickens, turkeys, swine
Oxytetracycline*	Chickens, turkeys, swine, cattle, sheep
Penicillin*	Chickens, turkeys, swine
Salinomycin	Chickens, cattle
Streptomycin*	Chickens, turkeys, swine
Tylosin	Chickens, swine, cattle
Virginiamycin	Chickens, swine
Sulfonamides*	Chickens, turkeys, swine

----

* Monitored Routinely by FSIS/USDA - Includes tetracycline and gentamicin

can result from the lack of adherence to withdrawal times, adherence to withdrawal times, illegal use of drugs, extra-label use of drugs, contamination of feeds, incorrect supplementation of feeds and other ways that have not been mentioned.

Violative residues are only one aspect of the total residue picture; residues can be within tolerance levels or they can be above legal limits. A potential problem exists if a residue is found in a large number of samples, even if only a small percentage of the samples contain violative levels of the drug. Similarly, a problem exists if many samples contain more than one residue, although few samples are at violative levels. An ideal situation would be a minimum number of positive samples with few, if any, at violative levels.

The inordinate frequency of appearance of residues of a specific drug may be due to misuse of the drug, or may instead be due to the appearance of new and more sensitive methodologies for the detection of the drug. Some areas of the food supply are screened infrequently or not at all, due to the lack of good analytical procedures for the task. The methods available may lack sensitivity and may be less precise or accurate than desired. In addition, many drugs were cleared for use at a time when residues of antibiotics and antimicrobial drugs were not considered to be a significant problem. However, since questions remain unanswered concerning the biological significance of residues and combinations of residues, their importance cannot be assessed accurately.

MONITORING AND SURVEILLANCE APPROACHES

The Food Safety Inspection Service (FSIS) of the U.S. Department of Agriculture (USDA) is the agency responsible for ensuring that meat and poultry are safe and free of adulterating residues. A violative or adulterating residue is defined as one which exceeds the tolerances established by the FDA.

The national monitoring program [4] provides information annually on the occurrence of violative residues of compounds for which there are established limits [5]. The monitoring-sampling program is a statistically-based selection of random samples from apparently healthy animals, and is designed to detect a residue problem that affects a specific percentage of the animal population. The number of samples chosen provides a 95% probability of detecting one violation when 1% of the animal population is in violation. When a problem in a major species is suspected, a larger number of samples are assayed.

The surveillance program was established to control the movement of potentially adulterated products when there was some prior knowledge or suspicion of a high potential for violative residues. The following brief summary of the in-plant tests used will help in understanding the data to follow:

(1) CAST - Calf Antibiotic and Sulfonamide Test. This test is used in veal calves under 3 months of age and 150 lbs or less. This test requires no laboratory confirmation and any violation results in condemnation of the carcass.

(2) STOP - Swab Test on Premises. This test detects the presence of residues of antibiotics in kidney tissue and requires other confirmation assays before condemnation occurs.

(3) SOS - Sulfa on Site. This assay system is used to detect residues of sulfamethazine in swine urine. Laboratory confirmation of violative levels is required.

The data presented in the following tables were taken from the data published in the Domestic Residue Data Book, National Residue Program 1988-1989, by the Food Safety and Inspection Service of the U.S. Department of Agriculture [6].

FREQUENCY OF RESIDUES IN MONITORING PROGRAM IN 1988

Nine antibiotics are routinely screened for in the FSIS/USDA system; they are chlortetracycline, erythromycin, gentamicin, neomycin, oxytetracycline, penicillins, streptomycin, tetracycline and tylosin. Chloramphenicol was not one of the antibiotics in the routine screen, but was part of the monitoring program in cows and calves and is included in the overall data presented. For the purpose of focusing the discussion on product class or species, some data presented in separate categories by the FSIS/USDA were combined. Bulls, cows, heifers and steers were combined under cattle; calves, fancy calves, bob calves, non-formula calves and western calves were combined under calves; market hogs, boars and sows were listed under swine; chickens, young and mature, were listed as chickens; turkeys, young and mature, were called turkeys; sheep and lamb were listed as lambs.

Table 2 shows the incidence of violative residues in the monitoring program for 1988. The overall frequency of violative samples is low. Of the major species, (beef, swine and poultry) swine had the highest frequency and the greatest diversity of violative drugs; violative residues in cattle were considerably lower with fewer antibiotics being found. Poultry (turkeys and chickens) exhibited the lowest incidence of the major species, with a combined incidence of 0.1%. Lambs, which are considered a minor species, have an incidence of 0.59%.

TABLE 2. Incidence of Violative Antibiotic Residues in the 1988 Monitoring Program

----

Production Class	Monitoring Samples[1]		Violative %	Antibiotics[2]
	Tested	Violative		
Horses	305	3	0.98	g
Cattle	1228	4	0.33	c,g,e
Swine	1381	10	0.72	a,b,d,f,g
Calves[3]				
Chickens	409	0	0.00	
Turkeys	541	1	0.18	g
Lambs	337	2	0.59	h
Ducks	319	0	0.00	
Geese	75	0	0.00	
Goats	280	1	0.36	g
Total	4875	21	0.43	

----

(1) Target tissue: kidney

(2) chlortetracycline (a), erythromycin (b), gentamicin (c), neomycin (d), oxytetracycline (e), penicillins (f), streptomycin (g), tetracycline (h), tylosin (i), chloramphenicol (j)

(3) Surveillance samples only

Unfortunately, the data for 1988, unlike previous years, did not include the total incidence of residues, both violative and non-violative. However low the percentages, if the data is reflective of overall production, a considerable amount of product containing violative residues appeared in the marketplace.

As was stated before, the surveillance program was used when there was prior knowledge or suspicion of violative levels of antibiotics/antimicrobials. Such knowledge comes from experience and a familiarity with agricultural production practices. Veal production is usually associated with fairly extensive use of antibiotics antimicrobials; hence, the incorporation of calves into the surveillance program. Since swine production is also associated with extensive drug use, it is not unexpected that swine might also have a high incidence of residues.

Table 3 shows the results of the 1988 surveillance program. As expected, there is a somewhat higher incidence of violative residues than in the monitoring program. Cattle showed a distinctly higher incidence of violative residues in the surveillance data than was observed in the monitoring data. Calves represented a rather substantial percentage of violative samples. Chickens and turkeys showed approximately the same incidence as in the monitoring study. The antibiotics found in the violative samples of cattle, swine and calves covered a wide spectrum; in many cases, there were multiple residues.

With the exception of the data on cattle and calves, a relatively low incidence of violative samples was found. Again, this incidence percent must be applied to the overall species production to have a complete picture of its significance.

TABLE 3. Incidence of Violative Antibiotic Residues in the 1988 Surveillance Program

----

Production Class	Monitoring Samples[1] Tested	Violative	Violative %	Antibiotics[2]
Horses	32	0	0.00	
Cattle	339	41	12.09	b - i
Swine	756	11	1.46	a,e,g,h
Calves	3394	88	2.59	a,c - h
Chickens	533	1	0.19	a
Turkeys	451	1	0.22	g
Lambs	497	9	1.81	d,f,g.h
Ducks	2	0	0.00	
Goats	1	0	0.00	
Total	6005	151	2.51	

----

(1) Target tissue: kidney

(2) chlortetracycline (a), erythromycin (b), gentamicin (c), neomycin (d), oxytetracycline (e), penicillins (f), streptomycin (g), tetracycline (h), tylosin (i)

The results of in-plant surveillance are shown in Table 4 . There are more samples taken and one would expect a higher frequency of residues than found in the monitoring survey. The Calf Antibiotic and Sulfonamide Test (CAST) indicated that a small percentage of the veal calves, 1.80%, had antibiotic residue levels in violation. Although the percentage of violative animals is low, the number of animals screened, 168,210, indicates that relatively large numbers of animals containing violative residues do reach the consumer. The drug residues found in violation did not follow any pattern. The Swab Test on Premises (STOP) showed that cattle and swine had the greatest incidence of violative antibiotic residues. The STOP test data approximates that obtained using the CAST system. Excluding the data for goats and lambs because of the small number of animals in the sample, the data should be reasonably reflective of what the consumer finds in the market place. For cattle, the spectrum of antibiotic residues found covers all the compounds used in the screen. This infers a generalized problem of misuse of the drugs, and suggests that an educational program might be useful to help minimize the problem.

TABLE 4. Incidence of Violative Antibiotic Residues - 1988
In-Plant Surveillance Sampling

| In-Plant Test | Surveillance Samples[1] | | Violative Antibiotics[2] |
	Tested	Violative	%
CAST	168,210	3,034	1.80	
STOP				
Species				
Horses	552	8	1.45	d,e,f,g
Cattle	53,468	2,068	1.45	a - j
Swine	2,218	24	1.08	d - g
Goats	87	1	1.15	f,g
Lambs	240	7	2.92	e,f,g
Total	224,775	5,142	2.29	

(1) Target Tissue: kidney

(2) chlortetracycline (a), erythromycin (b), gemtamicin (c), neomycin (d), oxytetracycline (e), penicillins (f), streptomycin (g), tetracycline (h), tylosin (i), chloramphenicol (j)

INCIDENCE OF VIOLATIVE SULFONAMIDE RESIDUES - 1988 MONITORING AND SURVEILLANCE PROGRAMS

Sulfonamide residues in some species have been a problem for about 10-12 years, probably resulting from a general misuse of the drug, as in feeding to animals when the growth stimulation benefits are no longer operative or when the animal is beyond the normal disease-treatment period.

Table 5 shows the results of the FSIS 1988 monitoring program for sulfonamides. In the general monitoring program, swine show a greater incidence of residues than other production species. This is not unexpected, since sulfonamides are used extensively in swine production. Sulfamethazine and sulfadimethoxine are the two drugs found as violative residues.

The surveillance program would be expected to yield a higher incidence of violative sulfonamide residues, as is shown in Table 6. A relatively high frequency of violative residues is found in swine and in cattle; both are a function of the high level of sulfonamide use in production. The incidence of violations found in calves is not unreasonable considering the use of drugs in veal production. In swine, the primary violative drug is sulfamethazine. The high incidence found in turkeys probably is a result of prior knowledge of a misuse; it is not representative of the generally low incidence of violations in turkeys.

TABLE 5. Incidence of Violative Sulfonamide Residues - 1988 Monitoring Program

----

Production Class	Monitoring Samples[1] Tested	Violative	Violative %	Sulfonamide[2]
Horses	306	2	0.65	f
Cattle	944	1	0.11	f
Swine	1936	35	1.81	d,f
Calves[3]				
Chickens	944	1	0.11	d
Turkeys	833	4	0.48	d
Lambs	344	3	0.87	d,f
Goats	103	0	0.00	
Total	5410	46	0.85	

----

(1) Target tissue: liver

(2) Sulfabromomethazine (a), sulfachlorpyridazine (b), sulfadiazine (c), sulfadimethoxine (d), sulfaethoxypyridazine (e), sulfamethazine (f), sulfamethoxypyridazine (g), sulfapyridine (h), sulfaquinoxaline (i), sulfathiazole (j)
(3) Surveillance samples only

TABLE 6. Incidence of Violative Sulfonamide Residues - 1988
Surveillance Program

Production Class	Surveillance Samples[1]		Violative	Sulfonamides[2]
	Tested	Violative	%	
Horses	2	0	0.00	
Cattle	107	7	6.54	d,f
Swine	3015	171	5.67	d,f
Calves	1186	18	1.52	f,h,j
Chickens	15	0	0.00	
Turkeys	59	6	10.17	f
Lambs	2	0	0.00	
Total	4386	202	4.61	

(1) Target tissue: liver

(2) Sulfabromomethazine (a), sulfachlorpyridazine (b), sulfadiazine (c), sulfadimethoxine (d), sulfaethoxypyridazine (e), sulfamethazine (f),sulfamethoxypyridazine (g), sulfapyridine (h), sulfaquinoxaline (i), sulfathiazole (j).

Probably the best indicator of the impact of sulfonamide residues for the consumer is offered by the Sulfa-On-Site (SOS) program, Table 7. The data reflects residue levels in muscle tissue, the primary edible product of the meat processing plant. The residue violation frequency is relativey low, 0.45 %, and the sample size is reasonably high. This indicates that, in general, the product available to the public contains a relatively low incidence of sulfonamide residues.

TABLE 7. Incidence of Violative Sulfonamide Residues[1] - 1988
In-Plant Test Surveillance Sampling

In-Plant Test	Surveillance Samples[2]		Violative
	Tested	Violative	%
SOS	85,720	385	0.45

(1) All residues are sulfamethazine

(2) Target Tissue: muscle

TABLE 8. Incidence of Violative Antibiotic Residues in the 1989 Monotoring Program

Production Class	Monitoring Samples[1]		Violative %	Antibiotics[2]
	Tested	Violative		
Horses	306	3	0.98	a,e,f
Cattle	627	6	0.96	d,e,f
Swine	1913	6	0.31	f,g,h
Calves	3722	33	0.89	c,d,e,f,g
Chickens	597	0	0.00	
Turkeys	488	0	0.00	
Lambs	320	0	0.00	
Ducks	325	0	0.00	
Geese	69	1	1.45	f
Goats	287	3	1.05	f,g
Total	8654	52	0.60	

(1) Target tissue: kidney

(2) chlortetracycline (a), erythromycin (b), gentamicin (c), neomycin (d), oxytetracycline (e), penicillins (f), streptomycin (g), tetracycline (h), tylosin (i), chloramphenicol (j)

INCIDENCE OF VIOLATIVE RESIDUES IN THE 1989 MONITORING AND SURVEILLANCE PROGRAMS

Table 8 shows the incidence of violative antibiotic residues from the 1989 monitoring program. As was seen for the 1988 program, the incidence was low overall, with the poultry samples showing no violative residues. The violation rate for cattle and calves was slightly less than 1 %. There was no pattern to the drug residues in violation.

Table 9 shows the results of the surveillance program for antibioics. Again, cattle, swine and calves showed the greatest incidence of violative residues, with no violations found in poultry. Where violations were found, several drugs were involved, with no pattern as to the drug usage.

FREQUENCY OF VIOLATIVE ANTIBIOTIC RESIDUES IN THE 1989 IN-PLANT SURVEILLANCE PROGRAM

The in-plant screening programs using the CAST and STOP systems showed that 2.62 and 3.42% of the samples, respectively, were positive, Table 10. Again, there was no apparent pattern to the antibiotics found in violation. Although these numbers indicate that the vast majority of animal carcasses should not contain residues of antibiotics in excess of the established tolerances, the numbers are considerably higher than those of the general monitoring and surveillance programs. Because of the large population sampled, these percentages of violations are probably more reflective of the extent of the problem.

TABLE 9. Incidence of Violative Antibiotic Residues in the 1989
Surveillance Program

| Production Class | Monitoring Samples[1] | | Violative % | Antibiotics[2] |
	Tested	Violative		
Horses	32	0	0.00	
Cattle	683	47	6.88	c,d,e,f,g
Swine	506	13	2.57	c,d,g,
Calves	597	13	2.18	c,d,e,f
Chickens	12	0	0.00	
Turkeys	24	0	0.00	
Lambs	6	0	0.00	
Goats	9	0	0.00	
Total	1869	73	3.91	

----

(1) Target tissue: kidney
(2) chlortetracycline (a), erythromycin (b), gentamicin (c),
neomycin (d), oxytetracycline (e), penicillins (f), streptomycin
(g), tetracycline (h), tylosin (i) chloramphenicol (j)

In 1989, an "Intensified Stop" program was initiated that assayed
animals for both antibiotic and sulfonamide residues. These data, shown
in Table 10, indicate that a high percentage of the samples had more
than one residue. It is not illogical to assume that if one residue was
in excess there would be a reasonable probability that more than one
drug would be above established tolerances, since many of the drugs used
in swine, veal and beef production are used in combinations. Samples
containing multi-residues are probably more prevelent than any of the
published data indicate.

INCIDENCE OF VIOLATIVE SULFONAMIDE RESIDUES - 1989 MONITORING PROGRAM

The results of the 1989 monitoring program for sulfonamides
indicated an overall low incidence of sulfonamide residues in excess of
tolerances. Cattle, swine and calves showed consistently low levels of
violative sulfonamide residues, as shown in Table 11.

INCIDENCE OF VIOLATIVE SULFONAMIDE RESIDUES - 1989 SURVEILLANCE PROGRAM

The results of the surveillance program, Table 12, show potential
problem areas in cattle, swine and calves. The high incidence in swine
is somewhat discouraging because of the extensive educational efforts
made for the proper use of sulfonamides in swine production. Chickens
were essentially free of violative residues. Although turkeys showed a
higher than expected incidence of violation, this should not be
considered a potential problem, since the sample numbers are small and
the surveillance program focuses on producers with a history of drug
misue. Unlike 1988, the pattern of violative drug residues in 1989 did
not single out any specific drugs.

TABLE 10. Incidence of Violative Antibiotic Residues - 1989
In-Plant Surveillance Sampling

| In-Plant Test | Surveillance Samples[1] | | Violative Antibiotics[2] |
	Tested	Violative	%
CAST	175,427	4,599	2.62	
STOP				
Species				
Horses	675	2	0.29	f
Cattle	79,309	2,868	3.61	a - j
Swine	4499	28	0.62	a,c - g
Goats	41	1	2.43	g
Lambs	559	9	1.61	c,f,g,h
Total STOP	85,083	2,908	3.42	
Intensified STOP - Animals Assayed for Antibiotics and Sulfas				
Species				
Cattle	1,203	648	53.86	
Lamb	11	2	18.18	
Swine	31	8	25.81	
Calves	35	14	40.00	
Total Intensified STOP	1,280	672	52.50	
Total	261,790	8,179	3.12	

---

(1) Target Tissue: kidney

(2) chlortetracycline (a), erythromycin (b), gemtamicin (c), neomycin (d), oxytetracycline (e), penicillins (f), streptomycin (g), tetracycline (h), tylosin (i), chloramphenicol (j)

The best indication of the overall residue problem comes from the Sulfa-On-Site (SOS) program. In 1989, the muscle tissue of almost 117,000 carcasses showed a low incidence of violative residues, Table 13. This is a promising statistic; it indicates that the vast majority of carcasses marketed do not contain violative residues.

DRUG RESIDUES IN MILK

The Grade A Pasteurized Milk Ordinance is recognized by the industry and public health agencies as the standard for milk sanitation. Legislative responsibility for the safety of the milk supply resides in the hands of the Food and Drug Administration.

In January, 1988, Brady and Katz [7] reported the incidence of antibiotic and sulfonamide residues in milk purchased from supermarket

TABLE 11. Incidence of Violative Sulfonamide Residues - 1989 Monitoring Program

---

Production Class	Monitoring Samples[1]		Violative %	Sulfonamide[2]
	Tested	Violative		
Horses	302	1	0.33	d
Cattle	632	1	0.16	f,j
Swine	1936	35	1.81	f
Calves	1221	5	0.41	d,f,j
Chickens	922	0	0.00	
Turkeys	819	5	0.61	d
Lambs	342	7	2.05	d,f
Goats	108	0	0.00	
Geese	68	0	0.00	
Total	6350	54	0.85	

---

(1) Target tissue: liver

(2) Sulfabromomethazine (a), sulfachlorpyridazine (b), sulfadiazine (c), sulfadimethoxine (d), sulfaethoxypyridazine (e), sulfamethazine (f), sulfamethoxypyridazine (g),sulfapyridine (h), sulfaquinoxaline (i), sulfathiazole (j)

(3) Surveillance samples only

shelves in central NJ, New York City and eastern PA. Approximately 40% of the samples contained sulfonamide residues in excess of 12.5 ppb, and approximately 40% contained tetracycline residues. Other antibiotics found were members of the ß-lactam family, the streptomycin family, the macrolides and chloramphenicol. In March of 1988 [8], the FDA initiated a 10-city survey to determine the presence of sulfonamide residues in the retail milk supply, and found that 73% of the samples contained levels ranging from 0.8 to 40.3 ppb. Five of the 36 (13.9%) samples contained sulfamethazine in excess of the 10 ppb "unofficial concern level".

In October 1988, after a concerted educational effort to reduce the incidence of sulfonamide residues, the National Conference on Interstate Milk Shipment sent a questionnaire to all state regulatory laboratories asking for data on the incidence of sulfamethazine in their milk. FDA analyzed the data from the survey and concluded that 247 of 4887 samples, or 5.05 %, contained sulfamethazine residues; 54 of the 4887 samples, or 1.10 %, contained residues above the unofficial 10 ppb concern level [8].

On December 29, 1989, the Wall Street Journal [9] published the combined results of two surveys of the milk supply. The results of a 10-city survey indicated that 19 of the 50 retail milk samples contained sulfonamide residues betwen 5 and 10 ppb. The other survey, of milk sold in Washington D.C., indicated that 4 samples of the 20 taken had

TABLE 12. Incidence of Violative Sulfonamide Residues - 1989
Surveillance Program

Production Class	Surveillance Samples[1] Tested	Violative	Violative Sulfonamides[2] %	
Horses	6	0	0.00	
Cattle	186	7	3.76	d,f,j
Swine	1068	111	10.39	b,d,f,h
Calves	198	6	3.03	f,j
Chickens	9	0	0.00	
Turkeys	108	4	3.70	d,i
Lambs	1	0	0.00	
Goats	1	0	0.00	
Total	1577	128	8.12	

----

(1) Target tissue: liver

(2) Sulfabromomethazine (a), sulfachlorpyridazine (b), sulfadiazine (c), sulfadimethoxine (d), sulfaethoxypyridazine (e), sulfamethazine (f), sulfamethoxypyridazine (g), sulfapyridine (h), sulfaquinoxaline (i), sulfathiazole (j)

sulfonamide residues between 5 and 10 ppb. The FDA responded with its own survey and collected 5 samples from each of 14 cities and used the same methodology, the Charm II receptor assay, an official AOAC method [10]. The FDA found that 60 of the 70 samples, or 85.7 %, contained residues of sulfonamides; 11 of the 70 samples, or 15.7 %, contained residues at concentrations above 10 ppb [8].

FDA confirmatory testing, using HPLC and Mass Spectrometry, did not confirm the presence of sulfonamide drugs. In retrospect, this was not an unexpected result. The Charm II receptor assay detects all members of the sulfonamide family used in agriculture. It measures all the molecular structures as one contribution. Hence, if there are low-level contributions of several sulfonamides, which is not an unexpected possibility, they would appear as a positive sulfonamide result. The HPLC procedure is less than reliable under 10 ppb and the Mass Spectral procedure used cannot detect residues under 20-30 ppb (FDA release, February 5, 1990); each detects individual compounds separately. Hence, the fact that these residues were not confirmed is not unexpected.

On February 7-8, 1990, WCBS Channel 2 TV News broadcast the results of a survey of milk purchased from store shelves in the New York City metropolitan area (New York City, Northern NJ, and Southern Connecticut) [11]. Of the 50 samples taken and analyzed, 18 contained sulfonamides at concentrations of 5-10 ppb; 14 of the 18 samples positive by the Charm assay contained sulfamethazine, as determined by an ELISA assay system. Forty of the 50 samples contained tetracycline residues; all 40 of the tetracycline positives detected by Charm II were positive using a microbial assay system [12]. The range of concentrations of the tetracycline residues was 25 - 180 ppb.

The results of these surveys pose problems in interpretation. Unlike the statistically-based FSIS/USDA monitoring, surveillance and in-plant surveys, the milk surveys were "snapshots" of a point in time, using small sample sizes. There are no parallel statistically-based programs for the sampling of the milk supply. In addition, the milk samples were examined for only a few drugs. Fifty-three drugs have been approved by FDA for use in dairy animals; 25 drugs have been reported to be used in an extra-label fashion. Only a small number of drugs are looked for in the milk supply; of these, only 6 drugs have confirmatory procedures.

TABLE 13. Incidence of Violative Sulfonamide Residues[1] - 1989 In-Plant Test Surveillance Sampling

| In-Plant Test | Surveillance Samples[2] | | Violative |
	Tested	Violative	%
SOS	116,726	318	0.27

(1) All residues are sulfamethazine
(2) Target Tissue: muscle

Regardless of the rhetoric used and the passions generated during that intense period of argumentation, charge and countercharge surrounding the milk surveys, the extent of the drug residue problem in milk remains undefined. Extra-label use complicates the problem further. The use of "concern levels" rather than the actual approved tolerances confuses the public. Based on the aforementioned results, it is not beyond reasonable logic to assume that market milk contains low concentrations of residues, at frequencies that vary.

MICROBIOLOGICAL SIGNIFICANCE OF ANTIBIOTIC RESIDUES

One of the unanswered questions related to antibiotic residues is the question of whether or not such residues can select for resistant populations. There is a dearth of information on this subject in the literature [13,14]. One reason it is not resolved is that published sources inevitably use a single compound at different concentrations, use levels that exceed the legal residues many-fold, or look at intestinal populations after feeding incurred residues. Almost all the published data suffer from the fact that only single residues were studied.

In 1991, Brady and Katz [15] exposed a sensitive strain of Staphylococcus aureus to single antibiotics and combinations of antibiotics, all at the legal tolerance levels, for a period of 14 days, transferring daily. Changes in the minimum inhibitory concentration (MIC) were determined. Increases of more than one tube (two-fold range of concentration) in the MIC after exposure are considered significant.

Table 14 shows the results of exposing the Staphylococcus aureus strain to antibiotics, both individually and in combinations of 3. Most of the single exposures did not result in any observable change in the MIC. Neomycin increased the MIC to oxytetracycline; penicillin increased the MIC to sulfamethazine and penicillin; oxytetracycline increased the MIC to itself; tylosin increased the MIC to oxytetracycline; streptomycin increased the MIC to oxytetracycline. Virginiamycin, gentamicin and sufamethazine did not increase the MIC of any compound beyond the two-fold range.

The results obtained using the combinations indicated that an additive or synergistic result occurred, with a definite increase in the MIC. Of the total of 45 MIC determinations for the combinations used, there were thirteen increases in MIC that were greater than two-fold, and 6 showed a two-fold increase. An MIC increase for 19 of the possible 45 determinations suggests strongly that an additive/synergy effect results from exposures to 3 antibiotics at tolerance levels.

This is an area where considerable work must be done in the future. The data from the in-plant animal surveillance programs indicate the presence of multiple residues in animal carcasses; the data from the milk surveys indicate the presence of multiple residues in the milk supply. It is important to know with a certain reasonableness the significance of residues on the development of resistance in bacteria.

OVERVIEW

There is no doubt that a small percentage of the animals used for human consumption contain residues that exceed the legal tolerances for those drugs. Although the percentage is small, the large number of animals produced and sold indicates that significant quantities of meat consumed have drug residues exceeding legal tolerances. The quality of any of these surveillance programs is a function of the ability of the program to identify problem areas, and to work to minimize the incidence of residues.

To accomplish the task of monitoring the nation's food supply for unwanted antibiotic and drug residues, it will be imperative to have methodologies that are simple to use, sensitive, accurate, precise and cost-effective. Analytical systems that can accomplish this task may very well be based on nonclassical approaches such as those related to, or based upon, immunological principles.

Paper of the Journal Series New Jersey Agricultural Experiment Station, paper No F-02-01112-91 supported, in part, by state funds.

TABLE 14. Changes in MIC of S. aureus ATCC 9144 to Individual Drugs After Exposure to Antibiotics/Antimicrobials at Residue Levels

Antibiotic/ Antimicrobial	Neo	Otc	Gen	Smz	Tyl	Pen	Stp	Vir	Cmp
				ug or units/mL					
Expected MIC	0.8	0.1	0.4	25	0.8	0.02	6.3	0.2	3.2
Control (aver.)	0.7	0.1	0.4	25	0.8	0.02	5.5	0.25	2.4
Neo	1.6	0.8	0.2	25	0.8	0.02	1.6	0.2	3.2
Gen	0.8	0.1	0.4	25	0.8	0.02	6.3	0.2	3.2
Pen	0.8	0.1	0.4	200	0.8	0.10	12.5	0.4	3.2
Otc	0.8	0.4	0.4	25	0.8	0.025	12.5	0.2	3.2
Smz	1.6	0.2	0.4	50	0.8	0.02	6.3	0.2	3.2
Tyl	0.8	0.2	0.4	25	6.3	0.02	12.5	1.6	3.2
Stp	0.4	0.4	0.4	25	0.8	0.02	6.3	0.2	3.2
Vir	0.8	0.1	0.4	12.5	1.6	0.02	6.3	0.8	3.2
Neo-Smz-Otc	3.2	0.8	0.4	100	0.4	0.08	3.2	0.4	3.2
Gen-Smz-Tyl	0.8	0.1	0.4	25	1.6	0.02	3.2	0.4	3.2
Pen-Smz-Otc	0.8	0.8	0.4	100	6.3	0.04	6.3	0.8	3.2
Stp-Tyl-Otc	1.6	0.4	0.4	400	0.8	0.32	12.5	0.4	3.2
Smz-Vir-Pen	0.8	0.2	0.4	400	1.6	0.16	12.5	0.8	3.2

Neo = neomycin; Gen = gentamicin; Pen = penicillin; Otc = oxytetracycline; Smz = sulfamethazine; Tyl = tylosin; Stp = streptomycin; Vir = virginiamycin; Cmp = chloramphenicol.

REFERENCES

1.   H.L. DuPont and J.H. Steele, *Rev. Infect. Diseases* 9, 447 (1987).

2.   "Feed Additive Compendium", The Miller Publishing Company, Minneapolis, Minnesota (1989).

3.   R.C. Livingston. Antibiotic Residues in Food: Regulatory Aspects, *in* "Agricultural Uses of Antibiotics," W.A. Moats, ed., ACS Symposium Series 320 p. 128, Washington, D.C. (1986).

4.   B. Schwab, and J. Brown. The U.S. Department of Agriculture Meat and Poultry Antibiotic Testing Program, *in* "Agricultural Uses of Antibiotics", W.A. Moats, ed., ACS Symposium Series 320 p.137, Washington, D.C. (1986).

5.   J. Brown, ed., "Compound Evaluation and Analytical Capability 1990 Residue Program Plan," Food Safety Inspection Service, U.S. Department of Agriculture, Washington, D.C. (1990).

6.   J. Brown, ed., "Domestic Residue Data Book, National Residue Program 1988-1989," Food Safety Inspection Service, U.S. Department of Agriculture, Washington, D.C. (1990).

7.   M.S. Brady and S.E. Katz. *J. Food Protect.* 51, 8.

8.   *Food Safety and Quality, FDA Surveys Not Adequate to Demonstrate Safety of Milk Supply.* Resources, Community and Economic Development Division, Washington, D.C. GAO/RCED-91-26.

9.   B. Ingersoll. Wall Street Journal, December 29, 1989.

10. "Official Methods of Analysis," 15th Ed., Association of Official Anaytical Chemists, Arlington, VA. Sec.988.08. (1990).

11. A. Diaz. New York WCBS Channel 2 TV News, February 7-8, 1990.

12. M.S. Brady and S.E. Katz. J. Food Protect. 52, 192. (1989).

13. M.S. Brady, R.J. Strobel and S.E. Katz. *J. Assoc. Off. Anal. Chem.* 71, 295. (1988).

14. L. Rollins, D. Pocurull, and H.D. Mercer. *J. Dairy Sci.* 57, 944. (1974).

15. M. S. Brady, and S. E. Katz. *In vitro* Effect of Multiple Antibiotic/Antimicrobial Residues on Selection for Resistance in Bacteria. Accepted, *J. Assoc. of Off. Anal. Chemists.* (1991).

21

RADIOIMMUNOASSAY OF ANTIBIOTIC/DRUG RESIDUES

IN FOODS OF ANIMAL ORIGIN

S.J. Steiner and J.N. Harris

New Jersey Agricultural Experiment Station
Rutgers, The State University of New Jersey

ABSTRACT

Radioimmunoassays have been utilized for the detection and quantification of antibiotic/drug residues in food products of animal origin. This type of immunological assay is a specific and sensitive analytical technique. Radioimmunoassays have extended the capability of assays to detect microgram to nanogram per kilogram quantities of antibiotic/drug residues. Correlation of results obtained with radioimmunoassay to chromatographic methods has been demonstrated. This paper will review the development of radioimmunoassays for the detection and quantification of penicillin and chloramphenicol, as examples, and will discuss the application of the technique to food products of animal origin.

INTRODUCTION

Berson and Yalow in 1958 described the utilization of radiolabelled ligands for the detection of trace amounts of antigens.[1] Their experiments demonstrated the ability of a radiolabelled antigen, insulin, to competitively inhibit the binding of non-labelled insulin to antibody. Application of competitive inhibition with radiolabelled antigen to radioimmunoassay (RIA) enabled assays to quantitatively measure small molecules at low concentrations.[1]

The application of radioimmunoassay has been diverse and has included the quantitative determination of hormones and antibiotics/drugs in samples from human, agricultural, and veterinary sources. Modifications of the techniques associated with radioimmunoassay have resulted in the development of more sensitive and specific immunological methodologies and their application to new assays.[3]

Radioimmunoassays for the quantitative detection of small molecular antigens such as drugs and hormones, generally employ radiolabelled antigens and specific antibodies. A fixed amount

Analysis of Antibiotic Drug Residues in Food Products of Animal Origin
Edited by V.K. Agarwal, Plenum Press, New York, 1992

23

of radiolabelled antigen is added to the sample and following competitive binding to an antibody, the bound complex is separated. The level of radioactivity is determined in the bound complex. To determine the quantitative level of unlabelled antigen, inhibition of the binding reaction using known levels of unlabelled antigen is completed. Thus, the labelled antigen and the unlabelled antigen contained in the sample compete for binding sites on the antibody. The ratio between labelled molecules bound to the antibody to those not bound is determined. Fewer labelled molecules are bound to the antibody as the concentration of the unlabelled antigen increases. An inhibition curve is constructed in which the degree of inhibition is plotted against increasing concentrations of unlabelled antigen. The curve demonstrates an inverse, non-linear relationship between the degree of inhibition and the concentration of the unlabelled antigen.[4] Unknown levels of antigen can be calculated from the curve by using the degree of inhibition and correlating it to a known concentration of antigen.

Several characteristics associated with the components of radioimmunoassay are critical in the function of the immunoassay. These include, the specificity of the antibody, the purity of the antigen and labelled antigen, and the ability to separate the bound and unbound complexes.

Radioimmunoassay was utilized by Hoffman to monitor hormonal residues in edible animal products.[5] This work is representative of the earliest classical employment of radioimmunoassay for the quantitative immunological determination of hormone residues in body fluids. The observations of Hoffman are applicable to current radioimmunoassay methodology. The presence of compounds and the subsequent production of metabolites can be quantitatively determined by radioimmunoassay. Such compounds are generally present in concentrations which range from parts per billion (nanogram(s) per gram) to parts per trillion (picogram(s) per gram) levels.

The experiments conducted by Hoffman demonstrated the need for antibody that was highly specific for the compound assayed by radioimmunoassays. The need for antibody which possesses high affinity has been documented.[5] Both polyclonal and monoclonal antibodies have been utilized in radioimmunoassay. Assay specificity is related to the utilization of antibodies that have been made in response to immunization with ligands which elicit an immune response. Antibiotics and drugs are not capable of eliciting an immune response following immunization because they have molecular weights less than 5000. Thus, they must be coupled or conjugated to a protein which acts as a carrier substance for the induction of antibody formation.[6] Albumin is often chosen as a carrier protein for immunization. The choice of carrier, the method of conjugation, and the purification of the immunogen are important parameters to consider for the preparation of a highly specific antibody.

The labelling of ligands is generally accomplished by the replacement of a carbon or hydrogen atom by ^{14}C or ^{3}H, respectively. Examples of assays which incorporate ligands labelled with ^{14}C or ^{3}H have been documented.[1,4,6,7] Ligands also may be labelled by utilization of ^{125}I. The majority of radioimmunoassays that have been developed has utilized iodinated ligands. Selection of the radioactive label for radioimmunoassay is influenced by the stability of the physical and biological half-lives, energy emitted, decay products, particle radiation, availability of label, cost and ability of the atoms to be used as a label.[2] Alternative radioimmunoassay methodologies have included the radiolabelling of antibodies rather than antigens in immunoradiometric assays (IRMA). However, immunoradiometric assays are generally utilized for the quantitative determination of proteins.

The separation of bound and unbound labelled ligand has been studied by many investigators. Catty has discussed several methods which have been employed for the separation of bound and unbound labelled ligands.[1] The ability to separate these assay components must be consistent over the range of unknown samples assayed. Solubility of the bound and unbound labelled ligands must be established. In general, the bound complexes are removed from solution and the unbound labelled ligand is left in solution. Alternatively, unbound labelled ligand is adsorbed onto a solid material. Examples of separation methodology cited by Catty include: 1. chemical precipitation by ethanol, ammonium sulphate or polyethylene glycol 2. utilization of an antiglobulin reagent or 3. attachment of antibodies to a solid phase and the subsequent removal of unbound labelled ligand by a washing step.

Radioimmunoassays utilize antisera which has been "titrated" against the labelled ligand. The binding of the labelled ligand to increasing dilutions of antibody as measured by counts of radioactivity is plotted, and curves are constructed from which an antibody dilution is selected that binds approximately 50% of the labelled ligand. This dilution of antibody, when incorporated into the competitive assay, will result in the optimum sensitivity for the quantitative determination of unlabelled antigen.

The extraction of hormones, drugs, or antibiotics from samples may include the homogenization of the sample followed by extraction with organic solvents. Another method utilized for extraction employs partition followed by column chromatography.[5] Further purification of the sample may be required to eliminate unspecific interference with the antigen-antibody reaction. In general, the degree of sample separation and purification are related to the type of sample to be assayed and to the degree of potential interference by the sample in the assay.

The specificity of a radioimmunoassay is determined by the substitution of ligands which possess similar chemical structures or biological activity in the assay. The precision, accuracy, and reproducibility of the radioimmunoassay are determined by the addition of "known" amounts of radiolabelled compounds to samples which do not contain the antigen to be assayed. Various extraction procedures should be utilized to determine the presence of interfering substances and/or compounds in the radioimmunoassay. Elimination of interfering substances and/or compounds is essential for the establishment of reliable radioimmunoassays.[1,5] Reproducible and consistent recovery is necessary for the quantitative determination of antibiotics and drugs.

REVIEW

There is a paucity of references which discusses the

utilization of radioimmunoassay for the quantitative determination of antibiotic/drug residues in foods of animal origin. Several references have cited the utilization of the Charm Test, or a radio receptor assay (RRA) in which the antibody in the immunoassay is replaced with a binder or bacterial cells (<u>Bacillus sterothermophilus</u>) on which receptors for various antibiotics/drugs are located.[1,9] An example of this type of immunoassay for the quantitative determination of penicillin in milk is described by Weiblen and Spahr.[10] Quantitative levels of penicillin were determined following administration and withdrawal of the antibiotic. Fifty percent of the milk samples contained penicillin residues in one sample after withdrawal. Carry over of penicillin residues was observed in the milking equipment. The sensitivity of the Charm Test was determined to be approximately 0.008 IU penicillin G equivalent per ml.

In 1987, Grover described the rapid detection of antibiotic residues in milk using a modification of the Charm Test.[11] The assay was a radioimmunoassay that was capable of quantitatively detecting 0.005 unit of penicillin per ml milk. Carbon 14 labelled penicillin and a "binder" were incubated with milk samples. The presence of unlabelled penicillin in the milk samples inhibited the binding of the labelled penicillin to the binder.

Examples of the application of radioimmunoassay to the quantitative detection of penicilloyl groups in biological fluids, including milk, were described by Wal and Bories et al.[12,13] The radioimmunoassay included labelled penicillin in which the penicilloyl group was attached to bovine serum albumin which had been labelled with ^{125}I. Dilutions of unlabelled penicillin were made in milk, urine and serum. The detection limit of the radioimmunoassay was 50 picograms of benzyl penicillonic acid. No non-specific interference in the presence of the biological samples was demonstrated in the radioimmunoassay.

The application of radioimmunoassay to antibiotic residues related to the presence of chloramphenicol in foods has been extensively studied by Arnold, vom Berg, Doertz, Mallick and Somogyi.[14] In 1984, these investigators described the results of a "new" radioimmunoassay for the detection of chloramphenicol residues in foods of animal origin including porcine tissue (muscle), bovine milk and hen's eggs following the therapeutic administration of the antibiotic. The radioimmunoassay was specific for the quantitative determination of chloramphenicol and the metabolites of chloramphenicol which were similar in structure. The chemical structure of the metabolites differed from that of chloramphenicol in the acyl side chain. The radioimmunoassay had a sensitivity of one microgram of antibiotic per kilogram of animal product. The results obtained by the radioimmunoassay were correlated to those obtained by a gas chromatographic method.

In 1985, Arnold and Somogyi continued to study the trace analysis of chloramphenicol residues in eggs, milk and meat by comparing results obtained with radioimmunoassay and gas chromatography.[15] The radioimmunoassay utilized crude extracts which were separated by partition in the presence of organic solvents. The solvents were evaporated to dryness and the purified residual material was incubated with antisera and labelled chloramphenicol analog. Dextran-coated charcoal was used for separation of the unbound complexes. The bound complexes were counted and the concentration of chloramphenicol residues was determined from a standard curve.

The extracts utilized for the radioimmunoassay were further purified on C_{18} reverse phase disposable cartridges for gas chromatography. The chloramphenicol was derivatized and chromatographed. The quantitative determination of the chloramphenicol was calculated by an internal standard method that compared the peak height of chloramphenicol and its derivative compound to a calibration curve.

The recovery of chloramphenicol residues was determined to be 85% for egg and meat samples and greater than 95% for milk samples. The precision of the radioimmunoassay varied and was dependent on the concentration of the chloramphenicol residues and the type of sample from which the residues were extracted. The sensitivity of the radioimmunoassay was one microgram per kilogram with milk, three micrograms per kilogram with egg, and five micrograms per kilogram with meat.[15] Levels of residues obtained with radioimmunoassay were confirmed by gas chromatography.

In 1986, Hock and Liemann, developed a radioimmunoassay for the quantitative detection of chloramphenicol. Competitive inhibition between chloramphenicol labelled with ^{14}C and antibody was demonstrated. Standard curves were constructed for the determination of the concentration of chloramphenicol. A correlation of 0.999 was obtained.[16]

Freebairn and Crosby, in 1986, also studied the analysis of chloramphenicol residues in animal tissue by radioimmunoassay.[17] Polyclonal antibody to chloramphenicol was produced following conjugation of the compound to keyhole limpet hemocyanin. Chloramphenicol was radioactively labelled with ^{14}C. Tissue samples spiked with chloramphenicol were freeze-dried, defatted and homogenized. Following centrifugation, C_{18} columns were utilized to further purify the extracts and the isolated chloramphenicol was assayed. The bound fraction was separated from the unbound labelled chloramphenicol with high molecular weight dextran-coated charcoal.

The results of the radioimmunoassay demonstrated recoveries of 15 to 90% of the isolated chloramphenicol. The sensitivity of the radioimmunoassay was 0.2 nanogram of chloramphenicol per gram of meat. Variation in the recovery of chloramphenicol was theorized to be related to the fat content of the samples.

Subsequent experimentation by Knupp, Bugl-Krieckmann and Commichau in 1987, consisted of a method for the detection of chloramphenicol in animal tissues and eggs by high performance liquid chromatography. The quantitative detection of the residues was confirmed by radioimmunoassay.[18] The radioimmunoassay had a sensitivity of greater than or equal to one microgram per kilogram of sample.

More recently, in 1989, a critical study of reference standards for residue analysis of chloramphenicol in meat and milk was conducted by Balizs and Arnold.[19] The study included the utilization of radioimmunoassay and gas chromatography with electron capture detection and mass specific detection. Similar to previous experimentation conducted by Arnold and Somogyi, reference samples were prepared.[14,15] The concentration of chloramphenicol was one microgram per kilogram in milk and ten micrograms per kilogram in meat. Approximately 70% of the chloramphenicol contained in the samples was recovered by the assay. The stability of chloramphenicol in milk stored at different temperatures (-30°C, -50°C and -80°C) was studied over one year. The stability of chloramphenicol was decreased when the drug was present at higher levels and the sample was stored at -30°C. Moreover, the authors demonstrated that the quantitative determination of chloramphenicol in plasma could be utilized to predict the concentration of chloramphenicol in tissues.

SUMMARY

This paper has summarized the utilization of radioimmunoassay for the quantitative determination of antibiotic/drug residues in foods of animal origin. The high specificity and sensitivity of radioimmunoassays were discussed. Correlation of results obtained with radioimmunoassay to chromatographic techniques was documented.

The development of highly specific monoclonal antibodies has led to the modification of radioimmunoassay methodology. In spite of the high sensitivity and specificity of radioimmunoassay, alternative immunoassays such as enzyme-linked immunosorbent assays, have replaced some radioimmunoassays for reasons of safety and economy.[2] Thus, the employment of enzyme-linked immunosorbent assays for residue analysis may offer an explanation for the lack of references for the quantitative determination of antibiotic/drug residues in foods of animal origin by radioimmunoassay.

REFERENCES

1. S. A. Berson and R. S. Yalow, Isotopic Tracers in the Study of Diabetes, Adv. Biol. Med. Phys. 6:349 (1958).

2. D. Catty and G. Murphy, Immunoassays using radiolabels, in: "Antibodies Volume II- a practical approach," D. Catty, ed., IRL Press at Oxford University Press, Oxford, England (1989).

3. M. N. Clifford, The History of Immunoassays in Food Analysis, in: "Immunoassays in Food Analysis," B. A. Morris and M. N. Clifford, eds., Elsevier Applied Science, New York, New York, (1985).

4. J. P. Gosling, A Decade of Development in Immunoassay Methodology, Clin. Chem. 36:1408 (1990).

5. B. Hoffmann, Use of Radioimmunoassay for Monitoring Hormonal Residues in Edible Animal Products, J. Assoc. Off. Anal. Chem. 61:1263 (1978).

6. B. A. Morris, Principles of Immunology, in: "Immunoassays in Food Analysis," B. A. Morris and M. N. Clifford, eds., Elsevier Applied Science, New York, New York, (1985).

7. R. J. Mayer and J. H. Walker, Uses of Polyclonal Antisera, in: "Immunochemical Methods in Cell and Molecular Biology," Biological Techniques Series, Academic Press Limited, San Diego, California (1987).

8. J. R. Bishop and C. H. White, Antibiotic Residue Detection in Milk- A Review, J. of Food Protection 47:647 (1984).

9. Charm Sciences Inc. Charm II Test For Beta-lactams in Milk Using Tablet Reagents. Operator's Manual.

10. R. Weiblen and S. Spahr, Determinacao da persistencia de penicilina no leite atraves de ensaio radio-immunologico ("Charm Test"), Pesquia Veterinaria Brasileira 2:133 (1982).

11. S. Grover, V. K. Batish, and J. S. Yadav, Rapid Detection of Residual Antibiotics in Milk, Indian Dairyman 39:393 (1987).

12. J. M. Wal, G. Bories, S. Mamas and G. Bories, Radioimmunoassay of Penicilloyl Groups in Biological Fluids, FEBS Letters 57:9 (1975).

13. J. M. Wal and G. F. Bories, In Vitro Penicillin Aminolysis: Application to a Radioimmunoassay of Trace Amounts of Penicillin, Analytical Chemistry 114:263 (1981).

14. D. Arnold, D. vom Berg, A. K. Doertz, U. Mallick, and A. Somogyi, Radioimmunologische Bestimmung von Chloramphenicol- Ruckstanden in Muskulator, Milch and Eiern, Archiv. fuer Lebensmittel-hygiene 35:131 (1984).

15. D. Arnold, and A. Somogyi, Trace Analysis of Chloramphenicol Residues in Eggs, Milk, and Meat: Comparison of Gas Chromatography and Radioimmunoassay, J. Assoc. Off. Anal. Chem. 68:984 (1985).

16. C. Hock and F. Liemann, Chloramphenicol: Entwicklung und exemplarische Anwendung eines Radioimmunoassays, Lebensmittelchem. Gerichtl. Chem. 40:63 (1986).

17. K. W. Freebairn and N. T. Crosby, The Analysis of Chloramphenicol Residues in Animal Tissue by RIA, in: "Immunoassays for Veterinary and Food Analysis," B. A. Morris and M. N. Clifford and R. Jackman, eds., Elsevier Applied Science, New York, New York, (1988).

18. G. Knupp, G. Bugl-Kreickmann, and C. Commichau, Ein Verfahren zur Absicherung von Chloramphenicol-Nachweisen in tier-ischem Gewebe and Eiern durch Hochleistungflussigchromato-graphie mit radioimmunologischem Nachweis (HPLC-RIA), Zeitschrift fuer Lebensmittel-Untersuchung und Forschung 184:390 (1987).

19. G. Balizs and D. Arnold, Reference Study for Residue Analysis of Chloramphenicol in Meat and Milk: a Critical Study, Chromatographia 27:489 (1989).

Current Problems Associated with the Detection of

Antibiotic/Drug Residues in Milk and Other Food

Stanley E. Charm

Charm Sciences Inc.
36 Franklin Street
Malden, MA   02148

INTRODUCTION

The focus of testing for antibiotic residues in milk is changing from preventing manufacturing problems and satisfying older regulatory requirements to more interest in food safety.

In 1981 the three hour B.stearothermophilus disc assay (BST) became the regulatory test for detecting antibiotic residues in milk. Although this test is particularly sensitive to beta-lactam drugs it is relatively insensitive to a number of commonly used antibiotics, e.g. (1), (2). However, beta-lactam drugs were a major problem in manufacturing cultured products and the test was widely used.

At this same time the ten minute microbial receptor assay (MRA) for beta-lactams was developed and commercialized as the Charm Test (3). The speed and sensitivity of the MRA permitted the testing of milk tankers before they unloaded at the processing plant (4).

For the first time, a processing plant could control and prevent beta-lactam contaminated milk from entering the plant. A contaminated tanker could be sent away and the milk would be delivered to another plant that didn't test. Usually it was possible to identify the producer causing the contamination, since a sample of each producer's milk making up the tanker accompanied the load. Frequently a penalty was imposed on the responsible producer.

The question of what to do with contaminated milk became more prominent as more plants tested milk tankers before unloading. It could be diluted or commingled (illegally), dried and used for animal feed, or dumped somewhere. A process was

Analysis of Antibiotic Drug Residues in Food Products of Animal Origin
Edited by V.K. Agarwal, Plenum Press, New York, 1992

developed using charcoal to remove residues as is done with water. After passage through the column the processed contaminated milk compares with antibiotic free milk (5). The process is under consideration by F.D.A as a possible salvaging procedure for contaminated milk.

In 1984-1985, the MRA was further developed to test for antibiotics beyond beta-lactams to include tetracyclines, sulfa drugs, aminoglycosides, chloramphenicol, novobiocin and macrolides. The BST is relatively insensitive to these additional drug families (6). The extended MRA was referred to as Charm II Test (CTII).

One event resulting from the CTII test was the discovery of widespread contamination in milk with sulfa drugs in 1986-1987 (7), (9).

This contamination was caused by the common use of over the counter sulfa drugs by producers and "extra label" use of sulfa drugs by veterinarians. Sulfa drugs are inexpensive and effective. Initially there was some attempt to control this contamination by testing at the plant, but most interest focused at educating producers about residues.

In December 1989, the Wall Street Journal and the Center for Science in the Public Interest carried out a national survey for antimicrobial drugs in milk using the MRA and found a significant incidence of low level sulfa drugs in milk (9). The F.D.A. carried out a similar study with CTII and found a similar sulfa drug incidence and finally confirmed the low level (less than 5 ppb) sulfa positives with HPLC (10). The sulfa drugs identified in the F.D.A. study starting with the highest incidence were sulfamerazine, sulfadiazine, sulfamethazine, sulfachloropyridazine, sulfadimethoxine and sulfamethiazole.

This activity lead to a Congressional subcommittee hearing, and a government accounting office examination of F.D.A. practices in drug monitoring (11). The F.D.A. has now recommended a change in regulatory testing beyond the BST in order to recognize other residues and has developed a list of prosecutorial concentrations for various drugs (12). A national program has been implemented to collect five samples per week (tanker milk or farm milk), to be tested for sulfonamides, tetracycline and chloramphenicol.

A COMPARISON OF VARIOUS TESTS WITH A MODIFIED MRA (CHARM II TEST)

A comparison of various tests used by the dairy industry for monitoring antibiotic residues was carried out by Virginia Polytechnic Institute (VPI) (2). This comparison was particularly useful in that it was carried out with the same representative drugs by the same operators (see Table 1). The CTII has the broadest spectrum with respect to F.D.A. safe or prosecutorial levels. It can detect more "safe levels" than other tests available (see Table 2).

TABLE 1. Minimum Tested Levels of Detection* (ppb):
Adapted from Virginia Polytechnic Institute Study (Reference 2)

Antimicrobial Drug	Safe/ Tolerance	Charm II Test	Charm Cowside II	CITE Test	Signal/ EZ Screen	Idetek (LacTek)	Penzyme III	Charm Farm Test	BR Test	Delvo Test P	Disc Assay
Penicillin G	10/0	2.5	5	5		5	5	2.5	10	2.5	5
Cephapirin	20/0	5	5	5		10	5	20	10	10	10
Cloxacillin	10/0	20	50	100		10	20	50	100	50	50
Ceftiofur	+/0	5	10	10		>100	50	50	100	50	100
Ampicillin	10/0	5	10	10		10	10	5	10	10	10
Amoxicillin	10/0	5	10	10		5	5	5	5	10	50
Tetracycline	80/	5		30				50	1000	420	>1000
Chlortetracycline	30/0	5		30				150	>1000	420	>1000
Oxytetracycline	30/	30		30				80	1000	200	1000
Erythromycin	50/0	50						1000	1000	400	1000
Tylosin	50/	150						50	50	100	1000
Sulfamethazine	10/	5	10	5	10	10		20	1000	>1000	>1000
Sulfadimethoxine	10/10	5	10	10	5	100		10	100	>1000	>1000
Sulfamerazine	10/	5	5		>1000	100		100	1000	>1000	>1000
Sulfathiazole	10/	5	100	10	>1000	>1000		100	1000	>1000	>1000
Sulfadiazine	10/	5	5		>1000	>1000		10	1000	>1000	>1000
Bovine Triple-Sulfa: Sulfapyridine	10/										
Sulfamethazine	10/	5	100	100	100	100		10	1000	1000	>1000
Sulfathiazole	10/			100							
Poultry Triple-Sulfa: Sulfamerazine	10/										
Sulfamethazine	10/	5	100	100	100	100		100	100	1000	>1000
Sulfaquinoxaline	10/										
Novobiocin	/100	100						750	>1000	1000	>1000
Polymixin B	/							>1000	>1000	>1000	>1000
Gentamicin	30/	30		30	30	30		150	>500	150	500
Neomycin	150/	>500		>500	10	>500		150	>500	150	>500
Streptomycin	125/	10		>1000		>1000		>1000	>1000	>1000	>1000
Spectinomycin	/	30		>1000		>1000		>1000	>1000	>1000	>1000

* Min. Levels where at least 13/15 replicates tested positive   + 16 mm zone B.St. Disc Assay (Reference 30)

TABLE 2. Method Detection at "Safe" Levels: Adapted from Virginia Polytechnic Institute Study (Reference 2)

Charm II Test
Penicillin
Cephapirin
Ceftiofur
Ampicillin
Amoxicillin
Tetracycline
Chlortetracycline
Oxytetracycline
Erythromycin
Sulfamethazine
Sulfadimethoxine
Sulfamerazine
Sulfathiazole
Sulfadiazine
Novobiocin
Gentamicin
Streptomycin
Spectinomycin (30)

Charm Cowside Test
Penicillin
Cephapirin
Ceftiofur
Ampicillin
Amoxicillin
Sulfamethazine
Sulfadimethoxine
Sulfamerazine
Sulfadiazine

Charm Farm Test
Penicillin
Cephapirin
Ceftiofur
Ampicillin
Amoxicillin
Tetracycline
Tylosin
Sulfadimethoxine
Sulfadiazine
Neomycin

CITE Test
Penicillin
Cephapirin
Ceftiofur
Ampicillin
Amoxicillin
Tetracycline
Chlortetracycline
Oxytetracycline
Sulfamethazine
Sulfadimethoxine
Sulfathiazole
Gentamicin

Delvotest P
Penicillin
Cephapirin
Ceftiofur
Ampicillin
Amoxicillin
Neomycin

BR Test
Penicillin
Cephapirin
Ceftiofur
Ampicillin
Amoxicillin
Tylosin

Disc Assay
Penicillin
Cephapirin
Ceftiofur
Ampicillin

Penzyme
Penicillin
Cephapirin
Ceftiofur
Ampicillin
Amoxicillin

LacTek
Penicillin
Cephapirin
Cloxacillin
Ampicillin
Amoxicillin
Sulfamethazine
Gentamicin

EZ-Screen
Sulfamethazine
Sulfadimethoxine

Penzyme III
Penicillin
Cephapirin
Ceftiofur
Ampicillin
Amoxicillin

Signal
Sulfamethazine
Gentamicin
Neomycin

In this comparison, CTII used antibody binders for tetracyclines and chloramphenicol in contrast to microbial receptors. Among the inhibition tests that require about three hours, the Charm Farm Test (CFT) in this study meets more "safe levels" than other inhibition tests (see Table 2).

For cloxacillin, the Lactek test detects 93% positives at 10 ppb (safe level), while CTII detects 33% at 10 ppb with 100% at 20 ppb. However the Lactek is refractory to ceftiofur with 0% positives detected at 100 ppb. Table 2 does not distinguish between detecting close to "safe level" and being entirely refractory.

In addition to MRA and inhibition tests there are the antibody tests, e.g. Cite, Lactek, and the enzyme kinetic assay, the Penzyme test for beta-lactams. As a test for surveying milk samples for residues the MRA has the advantage of detecting antibiotic functional groups which confers the broad spectrum capability on this method.

PRINCIPLES OF THE MICROBIAL RECEPTOR ASSAY

The MRA depends on the binding sites (enzymes) in microbial cells to complex the functional groups of antibiotics (enzyme inhibitors). In the assay, a microorganism containing a variety of binding sites is added to the sample, a labeled antibiotic (e.g. with exempt quantities of C14 or H3) is also added to compete for the binding site with any antibiotic in the same drug family that is in the sample (5). For example, beta-lactam drugs bind to D-alanine carboxypeptidase on the cell wall. Other binding sites are found on ribsomes (see Figure 1). The assay is based on the amount of labeled drug that binds to the receptor. The more label bound to the microbial cell sites the less antibiotic in the sample. An example of typical results is shown in Table 3.

Inhibition tests also bind antibiotics to receptors. In this case, the result of antibiotic binding is measured by growth inhibition and lack of acid production. In some cases there is good correlation between the receptor assay and inhibition assays, for example the CFT. With the CFT for sulfonamides and beta-lactams where the initial binding and final mode of action take place on the same site, there is excellent correlation between receptor and inhibition assays (see Figure 2). On the other hand macrolides/lincomycin and aminoglycosides which have additional subsites (13), (14) give less correlation. However, variations rarely exceed 10 fold between receptor and inhibition assays, e.g. the CFT. In some cases weak binding of the antibiotic to the receptor limits the MRA sensitivity. When this happens an antibody binder may be better, especially in a drug family with few members. In the CTII, chloramphenicol and tetracyclines use antibody binders.

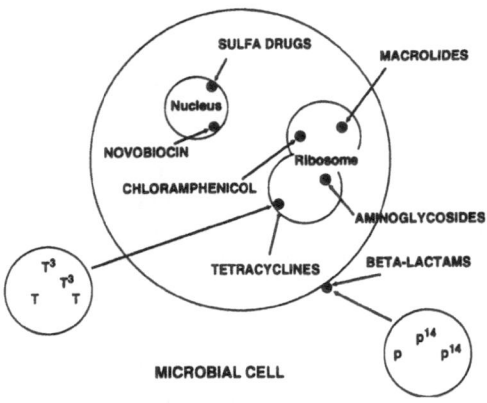

Charm Test Receptor Binding - Antimicrobial drugs in the sample compete with [^{14}C] or [^{3}H] labeled antimicrobials for specific sites in the microbial cell.

T = tetracycline	P = penicillin
T^3 = [^{3}H] tetracycline	P^{14} = [^{14}C] penicillin

---

* Although tetracyclines and chloramphenicol may be detected with a receptor assay, antibody binding is found to be superior in these assays, and has been incorporated in the Charm Test.

FIGURE 1. Charm Test Principles★

TABLE 3. Charm II Test for Sulfa Drugs in Milk Using Tablet Reagents: Typical Counts

Embossed Lot # SMTBL013   1 Minute Count on Charm II Scintillation Counter

Negative Control:		Positive Control:			
Zero	2204	10 ppb	837	Average:	877
	2554	Sulfamethazine	857	+ 15%:	132
	2238		835	Control Point:	1009
	2369		925		
	2392		905		
	2541		905		

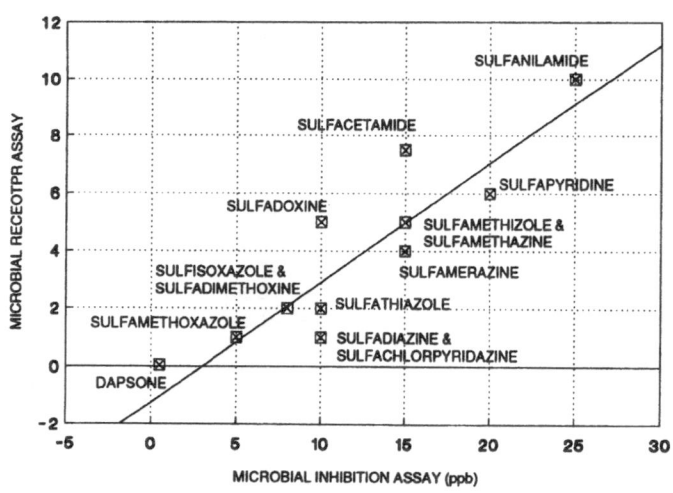

FIGURE 2. Microbial Receptor Assay (Charm II) vs. Microbial Inhibition Assay (CFT) for Sulfa Drugs

CONFIRMATION OF POSITIVE

Up to now, where testing was focused primarily on the beta-lactam drugs, the regulatory inhibition BST was often used to confirm MRA positives for beta-lactams. In most cases, the correlation is quite good between these two methods (as explained previously). Now that testing is going beyond beta-lactams, the scope and sensitivity of the MRA to other drugs is not matched by the BST.

For some drugs, antibody tests can be used to confirm receptor tests or the reverse, receptors confirm antibody tests, e.g., sulfamethazine, sulfadimethoxine, gentamicin antibody tests are available to confirm receptors.

A more broad spectrum inhibition test, the CFT may be used to confirm a receptor or antibody test provided the CFT has sufficient sensitivity, e.g. (see Table 4). The CFT is simple and takes about three hours.

HPLC methods are also frequently used to confirm and identify positives. The F.D.A. has such methods for eight sulfa drugs, tetracyclines, beta-lactams, chloramphenicol, novobiocin, available or under development (15).

An HPLC method developed with sufficient sensitivity to confirm the CTII is the HPLC-Receptorgram (16). One of its major advantages is the broad spectrum capability and sensitivity. In this method the CTII is used as an HPLC detector, thus identifying the retention time or fraction of the component causing a positive on the CTII. Retention times of standards of the drug family being analyzed are determined using an ultraviolet detector and correlated with the sample fraction retention time. An example is shown in Figure 3 of a sample testing positive with CTII for a tetracycline (antibody), confirmed with a tetracycline receptor assay and subjected to the HPLC-Receptorgram. It was identified as chlortetracycline (CTC). The CTII is then used to quantitate the CTC with the help of a CTC standard curve (in this case 25 ppb CTC).

The Receptorgram generally allows HPLC detection of individual drugs as low as the 1 to 5 ppb range. Drug quantitation may be carried out using HPLC, but is more convenient using the receptor assay or antibody assay with a standard curve of the identified drug. HPLC-Receptorgram methods exist for beta-lactams, tetracyclines, sulfa drugs, chloramphenicol and aminoglycosides.

It is also possible to show the HPLC fraction identified is an inhibitor by subjecting the positive HPLC fraction to an inhibition test sensitive to the drug. When this is done there are three different methods being brought to bear on the identification: a) identifying drug family, e.g. CTII; b) identifying specific drug, e.g. HPLC-Receptorgram; c) demonstrating the fraction containing the identified drug is an

TABLE 4. Limits of Detection (L.O.D.)★ in Milk: Charm Farm Test, Charm II Test and B.St. Disc Assay — Comparison of Various Antimicrobial Drugs

FAMILY GROUP	ANTIMICROBIAL DRUG✚	L.O.D.★ (ppb) CHARM FARM TEST	L.O.D.★ (ppb) CHARM II TEST	L.O.D.★ (ppb) B.St. DISC ASSAY✱
Beta-lactams	Penicillin	0.004 IU/ml	0.003 IU/ml	0.005 IU/ml
	Ampicillin	5	2.5	5
	Cephapirin	15	3	20
	Cloxacillin	40	3	35
	Oxacillin	30	3	30
Sulfa Drugs	Sulfamethazine	15	3	>1,000
	Sulfadimethoxine	10	5	>1,000
	Sulfathiazole	10	5	>1,000
	Sulfamethoxazole	10	10	>1,000
Aminoglycosides	Gentamicin	50	20	>1,000
	Neomycin	125	25▲	500
	Streptomycin	800	20	>2,000
Tetracyclines	Tetracycline	60	1	400
	Oxytetracycline	80	5	450
	Chlortetracycline	150	3	500
Macrolides	Erythromycin	400	15	>2,000
Novobiocin	Novobiocin	750	30	1,000

★ L.O.D. is defined as the lowest concentration significantly different from zero. To test at "safe level", refer to Charm II protocol ("Determination of Control Point" section). Charm II control point can be set for any level greater than L.O.D.

✚ The antimicrobial drugs listed are representative of their respective families. Other family members not listed are also detected.

▲ Neomycin sensitivity has been improved since VPI study (see Table 1).

✱ Data from IDF Bulletin 258 (1991).

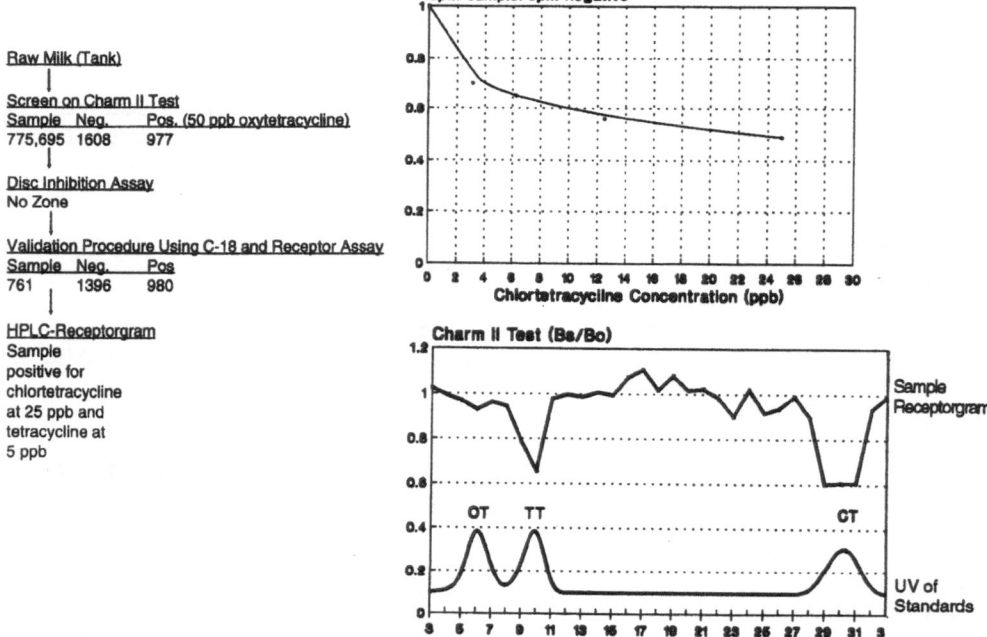

CHLORTETRACYCLINE STANDARD CURVE

Raw Milk (Tank)
|
Screen on Charm II Test
Sample Neg. Pos. (50 ppb oxytetracycline)
775,695 1608 977
|
Disc Inhibition Assay
No Zone
|
Validation Procedure Using C-18 and Receptor Assay
Sample Neg. Pos
761 1396 980
|
HPLC-Receptorgram
Sample
positive for
chlortetracycline
at 25 ppb and
tetracycline at
5 ppb

Figure 3. Example of Screening and Identification Process

inhibitor, e.g. CFT. This last step can only be accomplished when the inhibitor test is sufficiently sensitive to the identified drug. However, with larger sample volumes, the drug in a fraction can be concentrated before testing the fraction.

TESTING IN PROCESSING PLANTS

When the CTI was first introduced in 1978, an 18 hour B.subtilis disc assay detecting 12 ppb equivalent of Penicillin G was being used to test milk after it was mixed with other milk in the plant (17).

With a rapid test it is possible to prevent contaminated milk from entering and being owned by the plant. Rapid testing changed the equilibrium between buyers and sellers. A contaminated tanker had to find a plant that isn't testing. In 1981, regulations were changed so that greater than 3.5 ppb Pen G was a violation, i.e. a zone 16 mm or more with BST. This exacerbated the plant need to be sure penicillin type drugs were not present.

Other rapid tests are also now available for testing penicillin type drugs (see Table 1). These are antibody tests, e.g. Cite (18), and an enzyme kinetic assay, The Penzyme Test (19). These tests may not correlate as well as the MRA with inhibition tests for beta-lactams. For example, a comparison of Cite Test for beta-lactams showed it to be less sensitive than inhibition and microbial receptor tests (20). An excellent collection and comparison of antibiotic tests used throughout the world is shown in an International Dairy Federation monograph (21). However, the results are based on the developing laboratory or manufacturing company information, rather than a second party independent evaluation like the VPI study.

Since the discovery of contamination in milk from sulfa drugs, many processing plants test for these drugs. The MRA or CFT are the only methods that can detect all sulfa drugs in a single test. Antibody methods usually detect one sulfa drug at a time at low levels, e.g. Cite Antibody Test, three sulfa drugs, each in three separate tests. The F.D.A. prosecutorial level for one or all sulfa drugs is a total of 10 ppb in a sample (12), (22). The only rapid tests to accommodate this are the MRA methods.

It has been possible to rapidly eliminate sulfa drug contamination in an area by testing tankers. Individual producer samples on a contaminated tanker are checked and the producers contributing to the contamination can be determined. This usually warrants a visit to the producer from the company field man to correct the situation.

Tetracyclines have also come under scrutiny recently since widespread low level contamination has been noted from time to time in certain areas. Feed medicated with chlortetracycline

(CTC) and use of oxytetracycline (OTC), are the main reasons for this. CTC is allowed in feed at 70 mg/head/day (1400 lb cow), in continuous feeding (23). CTC has zero tolerance in milk, and OTC has no tolerance (24). However F.D.A. has set prosecutorial levels of 30 ppb for each of these. There are two rapid methods available for detecting CTC and OTC at close to these levels in milk, the CTII; 20 and 30 ppb respectively (3) and Cite test, 40 and 40 ppb respectively (18).

An incurred residue study for CTC in feed at two levels 70 and 140 mg/head/day is shown in Figure 4. Feeding at 70 leaves about 10 ppb residue, while 140 leaves 15 to 20 ppb in milk. In Figure 4, the control for the CTII test is set to accommodate the F.D.A. "safe level."

Medicated feed has the characteristic of producing wide spread contamination. It is legally possible to feed 140 mg/day/head (1400 cow), for 30 days, according to Code of Federal Regulations (CFR), (22). It is not clear what this really means since how much time should elapse between such feeding is not specified. In comparing the HPLC-Receptorgram for CTC in feed, and the CTC residue in milk, a conversion of CTC (feed) to tetracycline (T) in milk is observed (see Figure 5). The feed contains a ratio of CTC to T of 20:1 while the derived milk contains 4:1. The feed receptorgram has an unknown small peak with a tetracycline activity. This peak does not appear in milk.

It is also interesting to note that the HPLC-Receptorgram "activity" corresponds with the HPLC U.V. detection in feed. In milk the CTC and T are at too low a concentration to be detected well by U.V. and the Receptorgram "activity" is relied upon. (In the case of tetracyclines a sensitive cross-reacting antibody assay is employed rather than a microbial receptor for the Receptorgram.)

This points out the possibility of a high incidence of contamination in milk below "safe" or prosecutorial levels but greater than tolerance with CFR sanctioned treatments, such as with CTC. There is most likely a need to have a satisfactory contamination incidence level associated with a "safe level."

Testing milk tankers at the plant before unloading is the most advantageous point in the chain for exerting control. Here the volume of milk in a tanker justifies the cost of testing. Also the trained personnel and support at the plant level permit a sustained professional residue program.

Beta-lactam drugs have been used widely for sometime, yet the residue incidence for such use is less than 0.5%. The reasons for this are:

1. Milk withdrawal times are well known and adhered to.

2. Testing is routinely carried out at the plant level with rapid tests to screen tankers and also with inhibition tests after the fact.

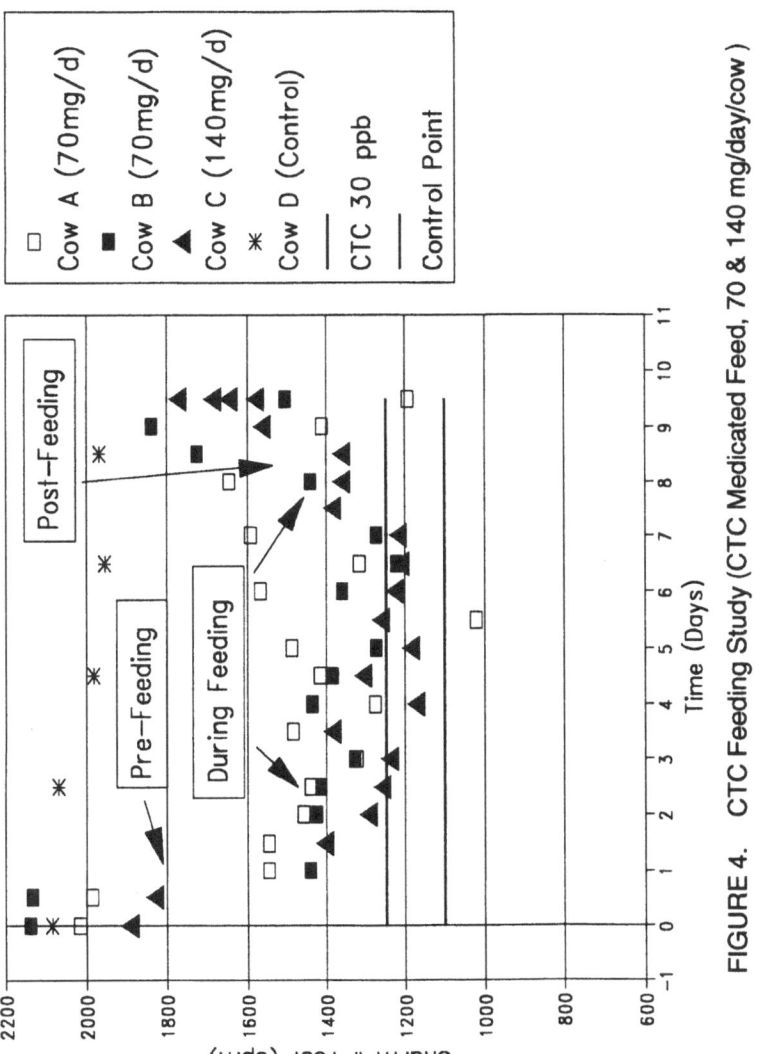

FIGURE 4. CTC Feeding Study (CTC Medicated Feed, 70 & 140 mg/day/cow)

41

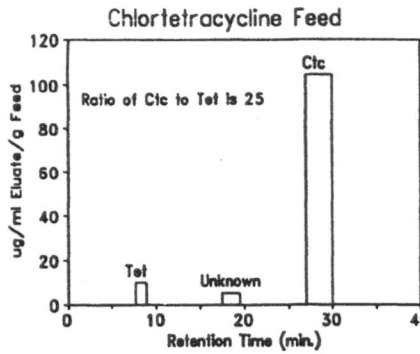

HPLC Receptorgram of Tetracycline in Medicated Feed: 1 g feed is extracted in 5 ml methanol and 20 µl portion is applied on HPLC using a Waters Nova-pak C18 column (5 um, 3.9 x 150 mm). Isocratic elution buffer contains 15% acetonitrile in a 5 mM ammonium phosphate, pH 3.8, and flow rate is 1 ml/minute. HPLC fractions were collected, dried and then assayed for tetracycline activity using the Charm II Test. Identification is achieved by correlating positive fractions with the retention times of tetracycline standards.

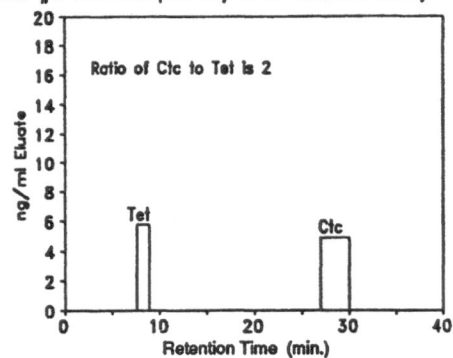

Milk Analysis: 20 ml sample is extracted, partially purified and dried. The concentrated sample is then subjected to HPLC analysis as above. Milk sample was obtained from dairy cow fed 70 mg of chlortetracycline in feed/day.

HPLC Chromatogram of Tetracycline Standards

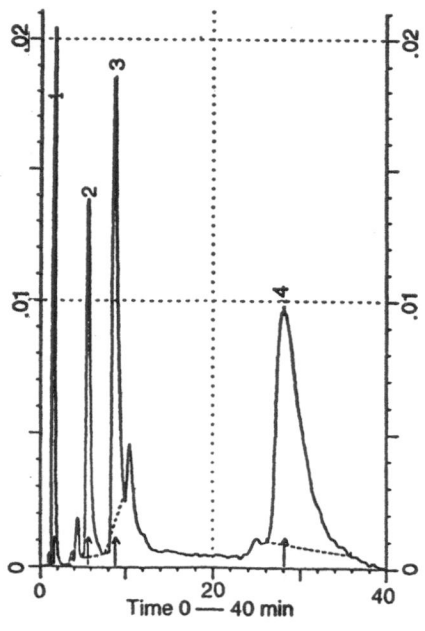

Standard Chromatogram: oxytetracycline, tetracycline and chlortetracycline were eluted at 5.7 min. (peak 2), 8.83 min. (peak 3) and 28.16 min. (peak 4), respectively.

FIGURE 5. HPLC Receptorgram for Chlortetracycline in Feed and Milk Showing Partial Conversion to Tetracycline

3. Testing maybe carried out at the farm for beta-lactams, using inhibition and rapid tests.

4. Producers frequently bring questionable milk to processing plants for testing before pick up at the farm.

FARM TESTING MILK FOR ANTIBIOTIC RESIDUES

There are several tests used for farm testing for beta-lactam residues, Delvotest (25), BR test (26), Charm Farm Test (3), Penzyme Test (19), Cite Probe Test (18). The Penzyme test and Cite test are rapid tests, 25 minutes and 15 minutes respectively while the former three are inhibition tests requiring 2.5 - 3 hours. It is estimated that 5 to 10% of producers use farm tests to monitor antibiotic treatment.

Of these farm tests, the CFT has the broadest spectrum for other residues, especially sulfa drugs (see Table 2). However the speed of the Cite test on the farm makes it useful for the drugs to which it can be applied, certain beta-lactams, tetracyclines, three sulfa drugs and gentamicin.

A scheme of inhibition tests for detecting various antibiotic residues in feed have been adapted to milk (27). Several years ago, this scheme identified low level tetracyclines in a market milk survey (28). Although this may have been greeted with some skepticism at the time, today it has increased credibility in view the results from other tetracycline tests, e.g. CTII, HPLC-Receptorgram, CFT, Cite (the latter two being used at the farm level).

Ceftiofur, a new beta-lactam veterinary drug, has recently been approved for use in lactating cows for intramuscular but not intramammary treatment. An intramuscular injection results in a residue less than 2 ppb equivalent of Pen G with the CTII, e.g. (see Figure 6). Preliminary study indicates the Charm II Test does not show positive when this drug is correctly used according to label. When it is positive, it is probable the drug was improperly used (29). If this low level residue remains for several days, successive daily injections may permit a residue accumulation to cause a positive, even though the drug is used according to label instructions. The F.D.A. has noted that a 16 mm zone on the BST regulatory test as a prosecutorial condition for this residue in milk from an intramuscular injection (30). All the tests for beta-lactam drugs used on the farm are capable of detecting ceftiofur residues at least at the prosecutorial level (see Table 2).

A REGULATORY TEST

The BST disc assay was the regulatory test used for the past ten years until the deficiencies in its spectrum and or sensitivity rendered it unsatisfactory as the single regulatory test. It was easy to use, inexpensive and not proprietary.

Today a regulatory test must at least satisfy F.D.A. prosecutorial levels, but with an eye to the CFR tolerances (see Table 5). There needs to be flexibility in this range. It should also be capable of detecting new drugs when they are introduced to the system. The known tests available today do not have all the characteristics of the ideal regulatory test. The CTII comes closest to fulfilling what a regulatory test should do.

The next best choice is the inhibition test CFT that fills in some of the gaps associated with the regulatory disc assay (see Table 2).

Both the CTII and the CFT have desirable characteristics for a regulatory test. The CTII is broad spectrum with high sensitivity. Several different reagents cover the spectrum of commonly used drugs. The CFT also has broad spectrum and sensitivity but less sensitive than CTII in some cases. However, CFT does wide spectrum testing in a single test.

A practical milk regulatory program might include, a) farm use of the CFT by producers who wish to use it; b) plant testing of milk tankers for certain drugs as needed but at least beta-lactams; c) positives could be checked with CFT.

Regulatory could primarily use the CFT with a Charm II as confirmatory when needed. All of these could be used in addition to the BST Disc Assay which is strongly rooted in regulatory testing. A regulatory test, with insufficient scope or sensitivity exposes milk to contamination since milk is bought and sold with the support of the regulatory test.

VERIFICATION AND COLLABORATIVE STUDIES

Any test used for regulatory must be verified and or certified to be used with confidence. The very best method is an A.O.A.C. Collaborative Study. A number of antibiotic tests in use today are A.O.A.C. Final Action tests, e.g. Charm I (4) (liquid reagents), Charm II (liquid reagents) (6), Delvotest (31), B.st disc assay (32).

A.O.A.C. is working to establish a system other than collaborative study to evaluate rapid tests. Consideration of R2 status, "Reviewed and Recognized" would not have official standing. The "Reviewed and Recognized" evaluation is required in order for a test to be used by a dairy. F.D.A. approval of R2 tests will also be a requirement for use in a dairy plant.

In 1991, the F.D.A. CVM carried out a Collaborative Study with nine laboratories and 55 coded samples including eight sulfa drugs using CTII. Although F.D.A. has not yet issued a report, an inspection of the data shows all samples greater than 5 ppb could be identified from negative or zero milk.

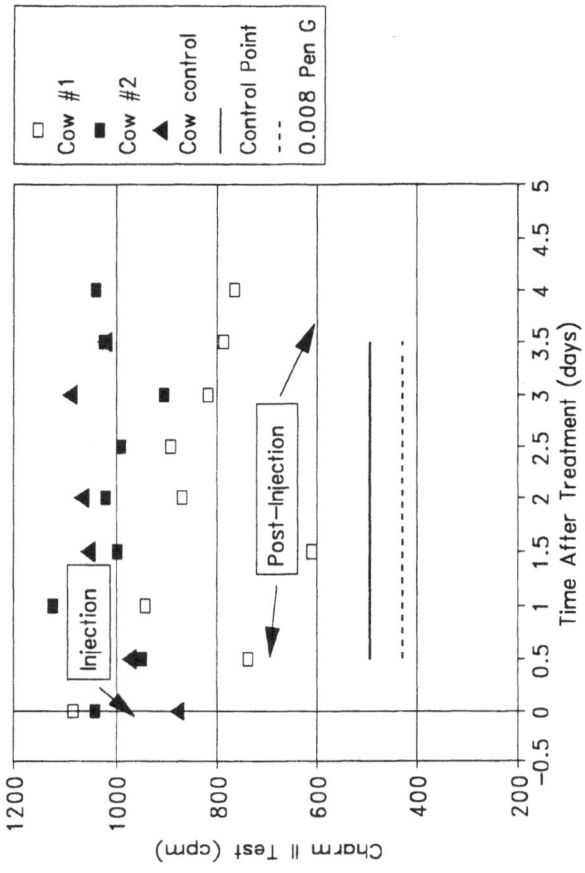

FIGURE 6. Ceftiofur Incurred Residue Study (Intramuscular Treatment at 750 mg/1500 lb)

45

TABLE 5. Summary of Antibiotic Residues — F.D.A. Tolerance or Action Levels★

COMPOUND	FAMILY	CFR REFERENCE	MILK✛	CATTLE	GOATS/ SHEEP	SWINE	POULTRY
Amoxicillin	Beta-lactams	21/556.38	10	10 Et	—	—	—
Ampicillin	Beta-lactams	21/556.40	10	10 Et	—	10 Et	—
Cephapirin	Beta-lactams	21/556.115	20	100 Et	—	—	—
Cloxacillin	Beta-lactams	21/556.165	10	10 Et	—	—	—
Penicillin	Beta-lactams	21/556.510	(4.8)	50 Et	0(40)Et	0(40)Et	0(40)Et
Sulfanilamide	Sulfa Drugs	FDA Draft	(10)	—	—	—	—
Sulfdiazine	Sulfa Drugs	FDA Draft	(10)	—	—	—	—
Sulfapyridine	Sulfa Drugs	FDA Draft	(10)	—	—	—	—
Sulfaquinoxaline	Sulfa Drugs	FDA Draft	(10)	—	—	—	—
Sulfachloropyrazine	Sulfa Drugs	21/556.625	(10)	—	—	—	0(100) Et
Sulfabromomethazine	Sulfa Drugs	21/556.620	(10)	100 Et	—	—	—
Sulfachlorpyridazine	Sulfa Drugs	21/556.630	(10)	100 Et	—	100 Et	—
Sulfadimethoxine	Sulfa Drugs	21/556.640	10	100 Et	—	—	100 Et
Sulfaethoxypyridazine	Sulfa Drugs	21/556/650	(10)	100 Et	—	0(100) Et	—
Sulfamethazine	Sulfa Drugs	21/556.670	(10)	100 Et	—	100 Et	100 Et
Sulfanitran and metabolites	Sulfa Drugs	21/556.680	(10)	—	—	—	0(100) Et
Sulfathiazole	Sulfa Drugs	21/556.690	(10)	—	—	100 Et	—
Chlortetraycline	Tetracyclines	21/556.150	(30)	100 M	100 M	1,000 M	1,000 M
Oxytetracycline	Tetracyclines	21/556.500	(30)	100 M	—	100 Et	1,000 M
Tetracycline HC1	Tetracyclines	21/556.720	(80)	250 Et	250 Et	250 Et	250 Et
Dihydrostreptomycin	Aminoglycosides	21/556.2000	0	0(500) Et	—	—	—
Streptomycin	Aminoglycosides	21/556.610	(150)	—	—	0(500) Et	0(500) Et
Gentamicin	Aminoglycosides	21/556.300	(30)	—	—	100 M	100 Et
Neomycin	Aminoglycosides	21/556.430	(150)	250 Et	—	—	—
Apramycin	Aminoglycosides	21/556.52	—	—	—	100 F	—
Erythromycin	Macrolides	21/556.230	0(50)	0(300) Et	—	100 Et	125 Et
Tylosin	Macrolides	21/556.740	50	200 M	—	200 M	200 M
Oleandomycin	Macrolides	21/556.480	—	—	—	150 Et	150 Et
Carbomycin	Macrolides	21/556.110	—	—	—	—	0(500) Et
Sulfomyxin	Polymixin	21/556.700	—	—	—	—	0(100) Et
Bacitracin	Bacitracin	21/556.70	20	500 Et	—	500 Et	500 Et
Chlorhexidine	Chlorhexidine	21/556.120	—	0(1)Et	—	—	—
Lincomycin	Lincomycin	21/556.360	150	—	—	100 Et	100 Et
Monensin	Monensin	21/556.420	—	50 Et	—	—	1,500 M
Spectinomycin	Spectiomycin	21/556.600	—	—	—	—	100 Et
Tiamulin	Multilin	21/556.738	—	400 L	—	—	—
8-a-hydroxy-mutilin	Multilin	21/556.738	—	3,600 M	—	—	—
Virginamycin	Virginamycin	21/556.750	—	—	—	100 M	100 M

★ All levels in ppb
✛ ( ) indicates Action Level

Key:
| Et = Edible tissue | M = Muscle |
| L = Liver | — = No tolerance |

Ideally a regulatory test will be collaboratively studied or be studied and approved by F.D.A. In practice individual state regulatory have some latitude in the testing procedures they employ. However the state regulatory test should at least meet the prosecutorial standards of F.D.A to prevent an unacceptable condition. To date, the VPI Study is the only comparison of the capability of various tests used for residue detection in milk (2).

TESTING FOR ANTIBIOTIC RESIDUES IN OTHER MATRICES
(Meat, Eggs, Fish, Serum and Urine)

The Charm Test and CFT have been adapted to testing for residues in other matrices beside milk.

With eggs, the CTII and CFT detect various commonly used drugs (see Table 6). Tolerances for eggs as described in the CFR (24) are noted in Table 7.

A 1990 survey of 70 market eggs using CTII for sulfonamides identified 20 percent low level positives (less than 50 ppb) using CTII. Two positives were identified as sulfachloropyridazine, and another as sulfamethazine using the HPLC-Receptorgram. The tolerance for sulfa drugs in eggs is 100 ppb. Although the USDA has a residue testing program for poultry, they do not have one listed for eggs (33).

The USDA program for meat includes the following tests (33):

1. The CAST Inhibition Test (calf antibiotic and sulfonamide test) is used only with bob veal calves (under 150 lbs. and less than 3 weeks old). CAST does not require laboratory confirmation of the result; any violation results in immediate condemnation of the calf.

2. The STOP Inhibition Test (swab test on premises) for antibiotic residues is used for a number of animal species. Laboratory confirmation is required.

3. The LAST Inhibition Test is used for living animals for urine or serum.

4. The SOS test, a chromatography test (sulfa on site) is used for detection of sulfamethazine residues in swine urine. The SOS is sensitive to 300 ppb. Laboratory confirmation of violations is required.

Laboratory confirmation is often carried with a variety of inhibition or specific antibody tests.

A comparison of the CFT with these other tests shows it is more sensitive for some drugs than the methods in use as well as the USDA FAST test (34) which is still under development. This

TABLE 6. Limits of Detection for Charm II Test and Charm Farm Test in Fresh Egg

DRUG FAMILY	ANTIMICROBIAL DRUG	L.O.D. (ppb)	
		CHARM II TEST	CHARM FARM TEST
Aminoglycosides	Streptomycin	150	3,200
Beta-lactams	Penicillin G	20	12
Chloramphenicol	Chloramphenicol	1	NA
Macrolides	Erythromycin	150	1,000
Sulfa Drugs	Sulfamethazine	25	80

TABLE 7. F.D.A. Tolerance Levels in Egg★

DRUG FAMILY	ANTIMICROBIAL DRUG	TOLERANCE (ppb)
Aminoglycosides	Streptomycin	0
Beta-lactams	Penicillin G	0
Macrolides	Erythromycin	25
	Tylosin	200
Sulfa Drugs	Sulfadimethoxine	100
	Sulfamethazine	100
	Sulfaquinoxaline	100
Tetracyclines	Chlortetracycline	0
	Hygromycin B	0

★ CFR Reference

is particularly true for sulfa drugs and tetracyclines (see Table 8). In reviewing the USDA residue program for tissues some confusion exists with respect to microbial inhibition methods. For example, the STOP test gentamicin limit of detection in a Canadian study (35) is 100 times greater than found by USDA (36). Similarly levels for neomycin, ampicillin, tylosin and oxytetracycline are 40, 16, 8, and 40 times higher (see Table 8). The company manufacturing the STOP test for USDA could not provide specific sensitivity information. There are apparently no studies confirming these tests in tissue with HPLC.

The CTII is more sensitive than CFT when compared for residues in muscle. Although the CTII is more sensitive and broad spectrum than the inhibition tests, the inhibition tests are more adaptable to meat inspector use at meat processing plants because of the little equipment needed and test time is not so critical. A comparison of LAST, CTII and CFT is shown in Table 9.

Serum and urine, are particularly useful for assessing the residue condition in live animals. The CAST and LAST are used by USDA for this purpose. The CFT for urine compared with the LAST test showed it is not as good for some drugs, e.g., streptomicin but better for others, e.g., gentamicin (see Table 9). The CFT requires only 2.5 hours vs. 18 hours for the LAST test.

A living animal contaminated with antibiotics may be held until the residue clears as indicated by testing serum or urine. A rapid urine or serum test for antibiotic assessment would be useful at the point where animals are purchased and sold. A comparison of the CFT with CTII when testing serum is shown in Table 10.

The SOS test is also commonly used in meat processing for checking sulfamethazine (33). This test employs methanol in plants without a hood causing the method to be re-examined.

The tolerances for antibiotic residues in fish are similar to other tissues. A list of approved drugs commonly used in aquaculture is shown in Table 11.

The CTII detection limits for fish are shown in Table 12. Studies are being carried at the Institute of Aquaculture, Stirling, Scotland on the application of the CTII for aquaculture (38).

SIGNIFICANCE OF ANTIBIOTIC RESIDUES IN MILK

There is a range of opinions about the significance of antibiotic residues in milk causing one of the most vexing problems. As with other types of residues, there is no way to know with certainty what the long term low level effect, if any, is, particularly on children.

The adverse effects known to be associated with high levels of antibiotic residues include; a) allergenic effects; b)

TABLE 8. Comparison of Various Tests for Antibiotic Residues in Tissue as Determined in Laboratories

		L.O.D. (ppm)					
DRUG FAMILY	ANTIMICROBIAL DRUG	CHARM II TEST	CHARM FARM TEST	FAST TEST [*]	CAST TEST [+]	STOP TEST [+]	STOP TEST [▲]
Aminoglycosides	Gentamycin	0.4	0.25	0.05	0.125	4.0	0.04
	Neomycin	0.8	0.3	0.1	0.06	8.0	0.25
	Streptomycin	0.15	1.5	1.0	0.2	1.0	0.25
Beta-lactams	Ampicillin	0.04	0.008	—	0.2	0.8	0.05
	Cephapirin	0.05	0.03	—	—	—	—
	Penicillin G	0.02	0.005	0.1	0.006	0.0125	0.0125
Macrolides	Erythromycin	0.4	0.3	0.05	0.15	0.8	0.025
	Tylosin	0.5	0.2	0.8	0.2	1.6	0.2
Sulfa Drugs	Sulfadimethoxine	0.01	0.06	4.0	—	—	—
	Sulfamethazine	0.025	0.1	3.0	0.2	—	—
	Sulfamethoxazole	0.01	0.06	—	—	—	—
	Sulfapyridine	0.05	0.2	—	—	—	—
	Sulfathiazole	0.01	0.06	5.0	—	—	—
Tetracyclines	Chlortetracycline	0.1	0.3	0.3	0.2	6.2	0.01
	Oxytetracycline	0.5	0.3	0.7	0.8	3.1	0.08
	Tetracycline	0.05	0.2	0.7	0.4	1.6	0.08

[*] Reference 34
[+] Reference 35
[▲] Reference 36

TABLE 9. Comparison of LAST Test,[*] Charm II Test and Charm Farm Test for Urine

		L.O.D (ppm)		
DRUG FAMILY	ANTIMICROBIAL DRUG	LAST TEST	CHARM II TEST	CHARM FARM TEST
Aminoglycosides	Gentamicin	1	0.08	0.5
	Streptomycin	5	0.08	20
Beta-lactams	Penicillin G	0.1	0.02	0.05
Macrolides	Erythromycin	1	0.2	2
Sulfa Drugs	Sulfamethazine	1	0.1	0.3
Tetracyclines	Tetracycline	1	0.1	1

[*] Live Animal Swab Test

TABLE 10. Comparison of Charm II Test and Charm Farm Test for Serum

DRUG FAMILY	ANTIMICROBIAL DRUG	L.O.D. (ppb)	
		CHARM II TEST	CHARM FARM TEST
Beta-lactams	Penicillin G	20	30
Macrolides	Tylosin	600	400
Sulfa Drugs	Sulfamethazine	40	150
Tetracyclines	Tetracycline	100	600

TABLE 11. Approved Antimicrobial Drugs Commonly Used in Aquaculture★

DRUG FAMILY	ANTIMICROBIAL DRUG	TOLERANCE LEVEL (ppm)
Folic Acid Analog	Ormethoprim	0.1 in salmonids & catfish
Sulfa Drugs	Sulfadimethoxine	0.1 in salmonids & catfish
	Sulfamerazine	0 in edible tissue of trout
Tetracyclines	Oxytetracycline	0.1 in salmonids, catfish & lobster
Other priority therapeutants pending approval:		
Aminoglycosides	Streptomycin	
Macrolides	Erythromycin	
Oxolinic	Oxolinic Acid	

★ Reference 36

TABLE 12: Charm II Test for Fish
Limits of Detection (L.O.D.)✚ for Antimicrobial Drugs

DRUG FAMILY	ANTIMICROBIAL DRUG	L.O.D. (ppb)
Aminoglycosides	Gentamicin	400
	Streptomycin	150
Beta-lactams	Ampicillin	40
	Cefotaxime	150
	Ceftiofur	150
	Cephalexin	100
	Cephapirin	50
	Cephradine	200
	Cloxacillin	150
	Hetacillin	50
	Nafcillin	250
	Penicillin	20
	Piperacillin	100
	Oxacillin	200
	Ticarcillin	500
Chloramphenicol	Chloramphenicol	5
Macrolides	Erythromycin	150
	Tylosin	500

DRUG FAMILY	ANTIMICROBIAL DRUG	L.O.D. (ppb)
Sulfa Drugs	Sulfacetamide	100
	Sulfachloropyridazine	10
	Sulfadiazine	50
	Sulfadimethoxine	10
	Sulfadoxine	5
	Sulfamerazine	20
	Sulfamethazine	25
	Sulfamethizole	30
	Sulfamethoxazole	10
	Sulfanilamide	75
	Sulfapyridine	50
	Sulfaquinoxaline	20
	Sulfathiazole	10
	Sulfisoxazole	10
Tetracyclines	Chlortetracycline	100
	Oxytetracycline	500
	Tetracycline	50

✚ L.O.D. is defined as the mean value of the matrix 3 standard deviations from the mean value of the matrix blank readings.

carcinogenicity; c) antibiotic resistant microorganisms. The "Linear No Threshold Model" often used to assess toxic effects and tolerances is thought to be conservative with the philosophy "better to be safe." Human body physiology is capable of accommodating small amounts of toxic substances, suggesting there is a satisfactory tolerance somewhere.

There are large populations in the world chronically exposed to antibiotic residues. Even the U.S. has been exposed to sulfa drugs in milk and meat in the past. There have not been acute or obvious effects nor has there been a focused study except for allergic response to beta-lactam residues in milk (39), (40), (41).

Obviously the food safety issue could be addressed with costly and lengthy toxicity studies for each drug in veterinary practice as was done in the case of sulfamethazine (42). A more practical approach is to control the residue through withdrawal times and appropriate regulatory programs.

SUMMARY

The testing for antibiotic residues reflects increasing interest in food safety.

A study by VPI supported by the dairy industry comparing the scope and sensitivity of the methods used showed the most comprehensive surveillance method available is the microbial receptor and antibody binder assay. With this method a wide spectrum of antibiotic families can be detected and identified.

Differences in scope and sensitivity among the various methods cause discrepancies in comparing sample results, i.e. negatives by one method are positive by another.

Although sanctioned by the CFR, CTC medicated feed for dairy cows may result in low level residue that is acceptable according to "safe level" standards. The frequency of these residues suggests that an incidence parameter possibly should be considered along with "safe levels."

The new beta-lactam drug ceftiofur has been approved without a milk withdrawal time for intramuscular injection. However the drug in some animals may leave a very low level residue which theoretically might accumulate with successive injections and cause a positive even though the drug is used correctly.

Confirmation methods are needed to support surveillance methods. The HPLC-Receptorgram is the most comprehensive HPLC method for identifying the most commonly used drugs. However, this has not received the official verification to bring it into widely accepted use.

New regulatory procedures for controlling residues in milk

are under consideration. These will require A.O.A.C. evaluation and F.D.A. approval. It is proposed to accomplish this by January 1992.

Methods for controlling residues in meat, egg and aquaculture fish are also being reviewed for improvement in sensitivity and scope. Different laboratories using STOP test report different results causing confusion.

Although the technology is available for effective control of antibiotic residues, these methods have to be acceptable to regulatory and F.D.A. In the context of residue control, antibiotic residues in milk are the easiest and most practical to control even though the question of significance lingers.

REFERENCES

1. J.T. Peeler, J.W. Messer, G.A. Houghtby, J.E. Leslie, J.E. Barnett. "Precision Parameters for A.O.A.C. Bacillus Stearothermophilus Disc Assay Based on F.D.A. Milk Laboratory Quality Assurance 1982-1986 Samples." J. of Food Protection, Vol 52, No. 12, Pages 867-870, (December 1989).

2. J.R. Bishop, S.E. Duncan, G.M. Jones, W.D. Whittier, 1991. "Evaluation of Animal Drug Residue Detection Methods." Report from Virginia Polytechnic Institute and State University, Blacksburg, VA 24061. Delivered at the NCIMS Meeting April 1991, Louisville, KY.

3. MFG by Charm Sciences Inc., 36 Franklin Street, Malden, MA 02148.

4. S.E. Charm and C.K Ruey, 1982. "Rapid Screening Assay for Beta-lactam Antibiotics in Milk: Collaborative Study." J. Assoc. Off Anal Chem, Vol 65, No. 5, P1186-1192, (MFG by Charm Sciences Inc., Malden, MA).

5. S.E. Charm, 1980. "Process for the Removal of Antibiotic from Milk." U.S. Patent 4,238,521.

6. S.E. Charm and C.K. Ruey, 1988. "Microbial Receptor Assay for Rapid Detection and Identification of Seven Families of Antimicrobial Drugs in Milk: Collaborative Study." J. Assoc. Off Anal Chem, Vol 71, No 2., P304-316.

7. S.E. Charm, E. Zomer, and R. Salter. "Confirmation of Widespread Contamination in Northeast U.S. Market Milk." J. of Food Protection, Vol 51, No. 12, P920-924, 1988.

8. G. Guest, 1988. "F.D.A. National Survey for Sulfamethazine in Milk." Statement at F.D.A. CVM Advisory Committee Meeting, April 12, Rockville, MD.

9. B. Ingersoll, 1989. "Milk is Found Tainted with a Range of Drugs Farmers Give Cattle." Wall Street Journal, Friday, December 29.

10. M.H. Thomas, 1990. "National Survey of Shelf Milk for Sulfonamides and Tetracyclines." Report of F.D.A. CVM April 3, Washington.

11. E.M. Zadjura, M.J. Rahl, M.L. Aguilar, W.M. Layden, S.W. Weaver, B.M. Karpman, 1990. "Food Safety and Quality - F.D.A. Surveys not Adequate to Demonstrate Safety of Milk Supply." G.A.O. - Report to the Chairman, Human Resources and Intergov. Relations Subcommittee. November U.S. General Accounting Office, P.O. Box 6015 - Gaithersburg, MD 20877.

12. S. Vaughn, 1991 National F.D.A. Statement at Committee Interstate Milk Shippers Conference, Louisville, KY, April 1991.

13. L.E. Bryan and H.M. VanDen Elzen. "Streptomycin Accumulation in Susceptible and Resistant Strains of Escherichia Coli and Pseudomonas Aeruginosa." Antimicrobial Agents Chemother 9:928,1976.

14. L.E. Bryan and H.M. Van Den Elzen. "Effects of Membrane Energy Mutations and Cations on Streptomicin and Gentamicin Accumulation by Bacteria: A Model for Entry of Streptomycin and Gentamicin in Susceptible and Resistant Bacteria." Antimicrob. Agents Chemother 12:163,1977.

15. M. Thomas, 1989. "Simultaneous Determination of Oxytetracycline, Tetracycline and Chlortetracycline in Milk by Liquid Chromatography." J. Assoc. Off Anal Chem, Vol 72, No. 4. And in F.D.A. Compliance Program Guidance Manual, 7371-008, TN 91-63, 6/17/91.

16. E. Zomer, S. Saul and S.E. Charm. "Confirmation and Identification of Sulfonamides in Milk by Liquid Chromatography with Microbial Receptor Assay." Submitted to Journal of Analyt. Chemists, August 1991.

17. ----------------------------------- 1985. "B. Subtilis Disc Assay in Standard Methods for Examination of Dairy Products." 15th Edition edited by G.H. Richardson. Published by American Public Health Association, Washington, D.C., P266-269.

18. ----------------------------------- 1989. Cite Test. MFG by Idexx Company, Portland, ME.

19. ----------------------------------- 1981. Penzyme Test. MFG by Smith-Kline Beecham Company, Philadelphia, PA.

20. G. Suhren and W.H. Heeschen, 1990. "Zum Nachneis Vol

B-Lactam Antibiotika in Milk Mit Cite-Test,
AgardiffusionsVerfahren und Mikrorobiellem Rezeptor Test."
Lebensmittelindustri and Milchwirtschaft, Vol 24,
P784-788.

21. W. H. Heeschen, 1991. "Detection of Antibiotics in Milk."
    Published by International Dairy Federation, Brussels,
    Belgium.

22. G. Guest, 1991. F.D.A. Center for Veterinary Medicine
    Advisory Committee Meeting, Bethesda, MD, April 10.

23. -------------------------------- 1988. "Chlortetracycline
    in Feed." In Code of Federal Regulations, Chapter 1,
    April 1. Published by Office of the Federal Register,
    General Services Administration, Washington, DC.

24. -------------------------------- 1984. "Antibiotic
    Tolerances." Code of Federal Regulations, Part 500-599.
    Published by Office of Federal Register, General Services
    Administration, Washington, DC.

25. "Delvotest." MFG by Giste Brocade, Netherlands.

26. "BR Test." MFG by Mueller Co., Germany.

27. S.M. Brady and S.E. Katz, 1989. "A Microbial Assay System
    for the Confirmation of Results of Receptor Assays for
    Antibiotic Residues in Milk." J. of Food Protection, Vol
    52, P198-201.

28. M.S. Brady and S.E. Katz, 1988. "Antibiotic/Antimicrobial
    Residues in Milk." J. of Food Protection, Vol 51, P8-11.

29. C. Miller, 1991. Personal Communication, Upjohn Co.,
    Kalamazoo, MI.

30. -------------------------------- 1991. "Milk Safety Branch
    Lists Tolerances/Safe Levels for 23 Drugs." Food Chemical
    News, Vol 33, P51-, July 26, 1991.

31. -------------------------------- 1983, Method 982.1. In
    Official Methods of Analysis 15th Edition, 1990
    Association of Official Analytical Chemists, Edited by
    Kenneth Helrich, 2200 Wilson Boulevard, Arlington, VA.

32. -------------------------------- 1984, Method 9821.17. In
    Official Methods of Analysis 15th Edition, 1990
    Association of Official Analytical Chemists, Edited by
    Kenneth Helrich, 2200 Wilson Boulevard, Arlington, VA.

33. Domestic Residue Data Book National Residue Program
    1988-1989. USDA Food Safety and Inspection Service,
    Jeffrey L. Brown, Editor 300 12th S.W., Washington, DC.
    20250.

34. S.A. Bright, S.L. Nickerson and N.H. Thaker, 1989. "FAST (Fast Antibiotic Sulfonamide Test)." Poster at the 103rd A.O.A.C. Annual International Meeting, St. Louis, MO.

35. G.O. Korsrud and J.D. MacNeil, 1988. "Evaluation of the Swab Test on Premises, The Calf Antibiotic and Sulfa Test and a Microbial Inhibitor Test with Standard Solutions of Antibiotics." J. of Food Protection, Vol 50, No. 1, P43-46.

36. R.W. Johnston, R.H. Reamer, E.W. Harris, H.G. Fugate and B. Schwab. "A New Screening Method for the Detection of Antibiotic Residues in Meat and Poultry Tissues." J. of Food Protection, Vol 44, No. 11, P828-831.

37. G.W. Fong and G.M. Brooks, 1989. "Regulation of Chemicals for Aquaculture Use." Food Technology, November, P92.

38. J.H. Brown, 1991. "Charm Enters the Prawn Unit." Aquaculture News, January, Vol 3.

39. J.M. Dendney, R.G. Edwards, 1984. "Penicillin Hypersensitivity - is Milk a Signigicant Hazard? A Review." J. Roy. Cox Med 78, P866-877.

40. B.B. Seigal, 1959. "Hidden Contacts with Penicillin." Bull World Health Organ, 21:703-713.

41. B.A. Friend, K.M. Shanani, 1983. "Antibiotics in Foods." In: Finley, J.W., Schwass, D.E., Editors "Xenobiotics in Foods and Feeds." Washington, D.C., American Chemical Society, P47-61.

42. Technical Report, March 1988. "Chronic Toxicity and Carcinogenicity Studies of Sulfamethazine in B6C3F1 Mice." National Center for Toxicological Research, Jefferson, AK 72079.

ELISA AND ITS APPLICATION FOR RESIDUE ANALYSIS OF ANTIBIOTICS AND DRUGS IN PRODUCTS OF ANIMAL ORIGIN

Deborah E. Dixon-Holland

Neogen Corporation
620 Lesher Place
Lansing, MI 48912

ABSTRACT

Development and application of enzyme-linked immunosorbent assays (ELISAs) for analysis of antibiotics and drugs used therapeutically and subtherapeutically in food producing animals have increased in the last decade. These immunochemical methods are capable of detecting low levels of residues in tissues as well as biological fluids (urine, blood, milk). These assays are rapid, sensitive, cost effective, require little sample clean-up and lend themselves to routine testing of large numbers of samples. They can be used for qualitative screening or quantitative analysis. The presentation will include a discussion of the principles of ELISAs. Examples of ELISAs for detection of specific drugs will be presented which include sulfonamides, chloramphenicol, B-lactams and aminoglycosides.

INTRODUCTION

The concern over the presence of drug residues in dairy and meat products has been manifested by the appearance of articles in publications such as Newsweek and The Wall Street Journal. Articles raised questions about milk safety, the sulfamethazine alert and the safety of the food supply (1-3). Routine testing of large numbers of samples is required to provide proper surveillance of the food supply. Therefore, the need exists for a method, that is precise and accurate, and enables testing of large numbers of samples on a regular basis. Enzyme-linked immunosorbent assays (ELISAs) exist or are being developed for detection of drug residues. These assays are attractive for a number of reasons which include: 1) Enzymes are used to label either antibodies or antigens, not radioisotopes. Therefore, the potential health hazards encountered when radioimmunoassays are used are eliminated. 2) These assays are rapid, and cost effective, 3) They can be used qualitatively to screen for the presence of an analyte above or below a fixed level, quantitatively to determine an actual concentration of the analyte in a sample based on a standard curve. These tests can find application in field situations (e.g. slaughterhouses) or in laboratories whether they be State or Federal government laboratories or private laboratories. They can be used to conduct basic research, to monitor the presence of one or more analytes in a sample, and can become an integral part of quality control programs.

Analysis of Antibiotic Drug Residues in Food Products of Animal Origin
Edited by V.K. Agarwal, Plenum Press, New York, 1992

57

Attention has been turned to ELISAs due to the drawbacks of current classical analytical methods for drug residues. Microbiological methods inhibition and diffusion assays have been used for decades (4,5). Chemical detection methods such as colorimetric methods (6,7) and chromatographic procedures (liquid chromatography (HPLC), gas chromatography (GC), thin layer chromatography (TLC) and mass spectroscopy (MS)) have been developed for drug residue analysis (8-11). These methods are extremely time consuming, are labor intensive (require extensive sample clean up and processing), may require expensive equipment, lack the specificity provided by antibody based tests and may encounter interferences from other drugs and sample components.

Either polyclonal or monoclonal antibodies can be produced for development of drug detection ELISAs (12). Polyclonal antibodies are commonly raised in rabbits as well as mice, sheep and goats. The term describes the products of a number of different cell types. Different clones respond to different antigenic determinants to create a mixture of antibodies. Monoclonal antibodies, on the other hand, are derived from a single clone and produce a homogeneous population of antibodies. Monoclonal antibodies are produced by hybridomas, fused spleen cells and myeloma cells. Rats and mice are used for spleen cell donors and myeloma cells (13).

The key tasks required for ELISA development are as follows: 1) A suitable immunogen must be prepared, 2) Immunize host animal (e.g. mouse or rabbit), 3) Obtain test bleeds to titer antisera for specific antibodies, 4) Develop an assay by optimizing (balancing) of antibodies and enzyme conjugate, 5) Apply the test to the desired sample matrix and, 6) Validate the method.

Since drug residues are low molecular weight molecules (<1000 Daltons), and are known as haptens, they must be chemically linked (in vitro or in vivo) to a carrier molecule, namely a protein, to elicit an immune response. If a reactive group, such as an amino group or carboxyl group is not present, then the drug must be derivitized to create a site for attachment to the carrier protein. Once the active group is added to the drug it is then chemically bound to the carrier protein. Commonly used proteins include bovine serum albumin (BSA), bovine gamma globulin (BGG) and keyhole limpet hemocyanin (KLH). Reports of the following conjugation methods have been reported in the literature: mixed anhydride reaction (14), carbodiimide (15), N-hydroxysuccinimide (16), periodate (17), glutaraldehyde (18-20) and diazotization (21). These methods quite simply, link the small molecule via an amino group to the amino group on lysine residues of the protein or via a carboxyl group to the amino group. The advantages or disadvantages of the specific reactions are documented in detail elsewhere in the literature (22).

Once a titer is observed in the test serum, and the reaction is shown to be specific, the reagents are optimized to generate the best possible curve in the desired range of analyte concentration. The reagents are often first standardized in pure solutions of the drug and then the desired sample matrix is applied to the assay. Samples which would be analyzed for any specific drug could be one or more of the following: 1) tissue (e.g. liver, kidney, muscle), 2) blood, 3) urine and 4) dairy products (e.g. milk, yogurt and cheese).

Once optimized and applied to the desired sample(s) the method must be validated. Generally, there are eight criteria which must be satisfied to complete the validation.

a.   Limit of detection (chemical sensitivity):. This is the smallest quantity or concentration of an analyte that can be reliably distinguished from background in the test.

b.   Crossreactivity (chemical specificity): The extent to which the assay responds to only the specified analyte and not to other compounds or substances in the sample.

c.   Reproducibility: The ability of the assay to duplicate results in repeat determination. It is the opposite of variability in the assay.

1.   Intra-assay variability: Variability between replicate determinations in the same assay.

2.   Inter-assay variability: Variability between replicate determinations from different groups.

d.   Reference correlation: The degree of closeness of the linear relationship between the results obtained using the ELISA and a reference assay (e.g. HPLC or TLC versus ELISA) over the range of the test.

e.   Sensitivity: The ability of the test to detect positive samples as positive. It is the percent positivity in a population of true positives.

f.   Specificity: The ability of the test to detect negative samples as negative. It is the percent test negativity in a population of true negatives.

g.   Overall accuracy: It is the combined or total ability of the test to correctly detect positive and negative samples (sensitivity and specificity = overall accuracy).

h.   Stability: It is the useable shelf life of the kit for specified storage conditions. It can be determined by accelerated aging studies and confirmed by real-time testing. This validation point, is, of course, important if a method is to be commercialized into a test kit format. An excellent discussion of method validation has been presented elsewhere (23).

It should be noted that ruggedness testing is often included in the validation. An example of ruggedness testing for an ELISA would be to determine test performance over a range of temperatures (e.g. extreme temperatures). This might be a concern when a test is used in the field in a setting exposed to seasonal temperature fluctuations (e.g. slaughterhouse).

ELISAs can be developed using microtiter wells (e.g. polystyrene) or membranes (e.g. nitrocellulose, nylon, polypropylene) as solid phases. The membranes are immobilized onto dipsticks, "cups" or "disks". Choices for enzymes include: horseradish peroxidase, alkaline phosphatase, glucose oxidase and B-O-galactosidase. Depending on the substrate chosen and the method used to terminate the reaction, different colors (e.g. green, blue and yellow) can be generated. Regardless of which enzyme is chosen it should satisfy a number of criteria: 1) High turnover number, 2) Easily detectable product (high extinction coefficient of product in a spectral region where substrate does not absorb light and if fluorescence detection is employed, high quantum yield of fluorescence of product), 3) Long term stability, 4) High retention of activity after coupling, 5) Absence of endogenous activity in sample, 6) Low cost and 7) Availability (24).

Table 1. Survey of Commercially Available ELISAs for
Detection of Drug Residues

B-lactams:	
amoxicillin	Idexx, Idetek
ampicillin	Idexx, Idetek
ceftiofur	Idexx
cephapirin	Idexx, Idetek
cloxacillin	Idexx, Idetek
hetacillin	Idexx, Idetek
nafcillin	Idetek
oxacillin	Idetek, Smith-Kline
penicillin	Idexx, Idetek, Smith-Kline
Tetracyclines:	
chloretracycline	Idexx
oxytetracycline	Idexx, Idetek
tetracycline	Idexx
Gentamicin	Idexx, EDI[a], Idetek, Smith-Kline
Neomycin	Smith-Kline
Sulfonamides:	
sulfadimethoxine	Idexx, EDI
sulfamethazine	Idexx, Neogen, Idetek, EDI, Smith-Kline, IDS[b]
sulfathiazole	Idexx
Tylosin	EDI

[a] Environmental Diagnostics Inc. (EDI)
[b] International Diagnostics Systems (IDS)

When developing an ELISA for a drug residue, either a competitive
direct ELISA or a competitive indirect ELISA can be developed. In the
competitive direct ELISA free hapten competes with an enzyme-labeled
hapten for a number of limited antibody sites attached onto a solid phase.
Unbound reactants are washed away before substrate is added. Following
substrate addition, the color produced is indirectly related to the amount
of hapten in the test sample. In the competitive indirect ELISA the
competition occurs between solid phase antigen and free antigen for the
antibody sites which are also free in solution. The antibody is not
immobilized as is the case in the competitive direct ELISA. The color
development is also inversely related to the amount of drug in the sample.

With each passing year there are increasing reports of immunoassays
for detection of drug residues in products of food producing animals.
There are numerous papers describing ELISAs for detection of B-lactams
(15). These will not be discussed here since other speakers have
described them in more detail. There are also numerous commercial kits
available for detection of antimicrobial agents including the B-lactams.
Table 1 shows a summary of ELISAs commercially available for detection of
antimicrobial residues in milk, urine, serum tissues and feeds. This list
was provided to producers of milk and dairy beef by the American
Veterinary Medical Association and National Milk Producers Federation.
They have implemented a quality assurance program to help producers
evaluate their production practices and thereby assure consumers that
dairy products and diary beef are of the highest quality possible. Use of
these rapid tests help producers control and prevent disease (e.g.
mastitis) in their herds, as well as prevent meat and milk from containing
violative drug residues (26).

Chloramphenicol is an antibiotic effective against a wide range of bacteria (e.g. gram negative infections in cattle). It is not approved for use in the USA but may be used surreptitiously. Concern over its presence in food products is due to the detrimental effects it may cause such as bone marrow depression, aplastic anemia and other serious blood disorders. Reports of ELISAs for its detection have been found in the literature (14,27).

In 1983 gentamicin was approved for use in swine as an injectable in 3 day old piglets for treatment of neonatal diarrhea and as an oral solution in drinking water to treat neonatal diarrhea or swine dysentery. A study was conducted on 3182 obtained from slaughterhouses sera using a competitive direct ELISA (17). The test was able to detect as little as 2.3 ng/ml gentamicin in serum, and 200-400 samples per day per analyst could be tested. This is a perfect example of how large numbers of samples can be tested using an ELISA. The gentamicin antibody was fairly specific for gentamicin. It cross-reacted 33% with sisomicin and 4% with netelmicin, but not with any drugs used agriculturally. The highest level of gentamicin detected in the samples was 130 ppb.

An in depth discussion of ELISAs for detection of two different drug residues follows.

Experiment

Sulfamethazine. A twenty minute ELISA was developed for detection of sulfamethazine in swine serum (28). Briefly, antibody was coated onto the surface of microtiter wells (100 mcl/well of a 1:2000 solution of antibody diluted in 2.5 mm Tris buffer, pH 8.0) according to manufacturer's specifications. Unbound antibody was removed following washing with water (300 mcl/well for five times). To perform the assay the caps were removed from a strip of eight minitubes. Each tube contained 0.4 ml of a 1:4000 dilution of enzyme conjugate (sulfamethazine horseradish peroxidase (SMZ-HRP)) diluted in a serum replacement medium, and 0.3 ml of 0.1 M phosphate buffered saline, PBS, pH 7.5. The PBS was a diluent for serum samples. Tube 1 contained 0.1 M of a 100 ppb serum control (low standard) and Tube 8 contained 0.1 ml of 400 ppb serum control (high standard). A blood

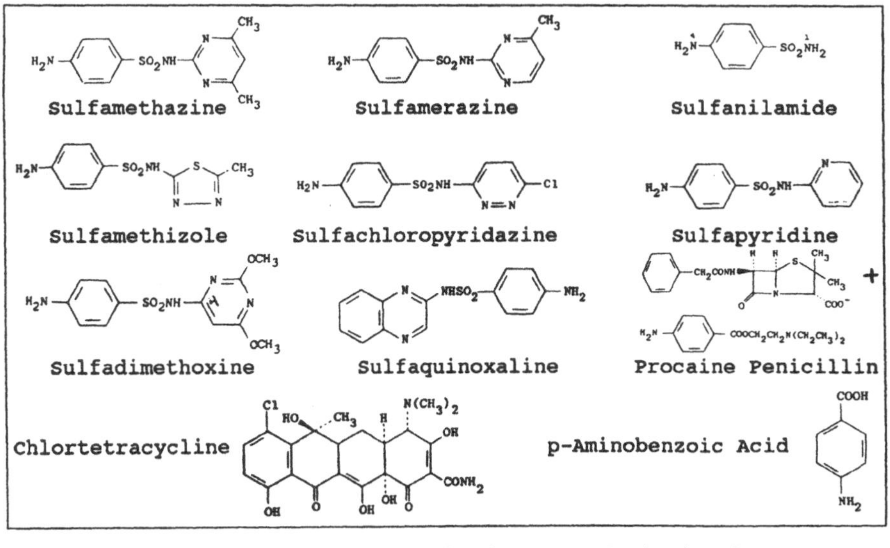

Figure 1. Sulfamethazine and Other Important Structures.

sample was collected at the time of slaughter from the heart. The blood was allowed to clot and the serum used for testing. Samples 1-6 were added to Tubes 2-7, respectively (0.1 ml/tube). Using an 8 channel pipettor the solutions were mixed in the minitubes and 0.1 ml of control or sample was added, using the 8 channel pipettor, to the appropriate antibody coated well. The reaction was carried out at room temperature for 10 minutes. The excess reagents were removed by washing (0.3 distilled water/well) the wells 5 times. Excess water was removed by tapping the inverted wells onto adsorbent paper. Next, using a dropper bottle, 2 drops of substrate (tetramethylbenzidine (TMB) and hydrogen peroxide (H_2O_2) in a 2:1 ratio) were added to each well. The reaction was carried out for 10 minutes at room temperature. The enzyme activity was inhibited by adding 2 drops of a stopping reagent containing a red dye. Results were read visually and using a color comparator (AgriScan, Neogen Corporation).

The ELISA was compared to TLC (29) to determine a reference correlation.

Avermectins (Ivermectin). A competitive indirect ELISA was developed for detection of ivermectin (16). Briefly Immulon 2 wells were coated with a limiting amount of ivermectin 1-conalbumin (overnight at 4°C) at a concentration of 25-50 ng/carrier protein 0.1 ml of coating buffer (PBS, pH 7.5). Blocking was done according to manufacturers instructions. Standards (0.1 ppb-1 ppm) or samples were mixed with the limiting dilution of antibody (PBS-Tween containing 10% (v/v) acetonitrile or other organic solvent). The mixture was incubated overnight at room temperature in tightly sealed polypropylene tubes. The coated wells were washed with PBS-Tween (3 washes with a wash bottle and excess water tapped out on adsorbent paper). The wells were blocked for 3 minutes at room temperature with PBS-Tween-1% bovine serum albumin (BSA). The wells were washed again and the solutions which had been incubated in the polypropylene tubes was added (0.1 ml/well) to the blocked wells containing solid phase antigen and the mixtures were incubated at room temperature for 2 hours. The wells were washed again and 0.1 ml of alkaline phosphatase conjugated goat antimouse antibody (1:1000 in PBS-Tween) was added to each well. Following another 2 hour incubation at room temperature the wells were washed and substrate was added (p-metrophenol phosphate (Sigma 104 substrate tablets)) 1 mg/ml in 10% (w/v) diethanolamine hydrochloride pH 9.8-0.4 mM $MgCl_2$-3 mM NaN_3). Color development was monitored on a Multiskan EIA reader (Flow Laboratories) interfaced with a Mackintosh computer, and the rates of the reaction (SA_{405}/min x 103) were calculated by linear regression.

Table 2. Lowest Detectable Limit of Sulfamethazine ELISA.

ppb	O.D.	CV
0	1.085	8.0
100	0.574	10.8
200	0.428	9.6
300	0.357	7.0
400	0.303	11.2
500	0.260	10.4

$$\text{Lowest Detectable Limit} = \frac{x_o - 2 * (S.D.)}{x_o}$$

$$= \frac{1.085 - 2(0.087)}{1.085} = 84\%$$

This corresponds to 13 ppb on a logit-log curve.

62

Compound	Concentration Required to Inhibit 50% of Antibody Biding (ppb)	Relative % Cross-Reactivity
sulfamethazine	90	100
sulfamerazine	630	14
sulfathiazole	>50,000	<0.18
sulfapyridine	>50,000	<0.18
sulfaquinoxaline	>50,000	<0.18
sulfachloropyridazine	>50,000	<0.18
sulfadimethoxine	>50,000	<0.18
sulfanilamide	>50,000	<0.18
chlortetracycline HCl	>50,000	<0.18
procaine penicillin	>50,000	<0.18
p-amino benzoic acid	>50,000	<0.18

[a]Relative % Cross-Reactivity =

$$\frac{\text{Concentration of Sulfamethazine Required to Inhibit 50\% of Antibody Binding (ppb)}}{\text{Concentration of Other Antimicrobial Agent Required to Inhibit 50\% of Antibody Binding (ppb)}} \times 100$$

RESULTS AND DISCUSSION

Sulfamethazine. The rapid ELISA was developed for detection of
sulfamethazine in swine serum. Description and discussion of this test
was included due to all the uproar and published reports of sulfonamide
violation in swine tissues. There is a direct correlation of how much
sulfamethazine is present in the serum versus the concentration. It has
been shown that a level of 100 ppb of sulfamethazine in the serum will
lead to 100 ppb in the liver and 25 ppb in the edible tissue. If the
serum level of sulfamethazine is 400 ppb then the level in the liver is
400 ppb, while the concentration in the edible tissue is 100 ppb. (30,31)
The FDA has established a tolerance level of 100 ppb in tissue and liver
(32). Being able to test for concentrations of the drug in the liver and
tissue without having to sample tissue is an advantage. The test can be
used on live during the 15 day withdrawal period to determine if the serum
level falls below the levels to insure that violative residues will not be
found at the time of slaughter and lead to carcass condemnation as well as
economic loss to the producer.

The limit of detection was determined to be 13 ppb (Table 2) which
is well below the concentration of the low control presented in the assay.
The antibody was found to be highly specific for sulfamethazine (Table 3).
There was a low percentage of crossreactivity to sulfamerazine. This is
not surprising since sulfamerazine lacks one methyl group that
sulfamethazine possesses (Figure 1). Other sulfonamides relatively
similar in structure were not recognized nor were penicillin and
chlortetracycline which are commonly added into swine feed with
sulfamethazine. The assay was determined to be highly repeatable with
1.78-4.97% intra-assay variability and 6.14-8.77% inter-assay variability.
(Table 4) The reader should keep in mind the Horwitz (33) overview of
relative error of analysis when dealing with parts per billion (ppb)
concentrations. It is not abnormal that the relative error be ±50% in ppb
range and ±25% in the parts per million (ppm) range. Considering that the
coefficients of variation were below 10% the procedure is both sensitive
and reproducible.

Table 4. Reproducibility of the ELISA.

Replicate	0	50	100	200	400	500	x
	ng/ml (absorbance at 650 nm)						
1A	1.098	.771	.638	.499	.497	.390	
1B	1.165	.783	.688	.521	.490	.385	
1C	1.162	.852	.706	.538	.487	.377	
1D	1.132	.818	.730	.553	.508	.378	
1E	1.161	.953	.790	.556	.503	.401	
1F	1.171	.903	.817	.588	.504	.397	
x	1.148	.847	.728	.542	.498	.388	
s	.028	.071	.066	.031	.008	.010	
C.V.	2.44	8.37	9.10	5.67	1.67	2.54	4.97
2A	1.368	.915	.764	.567	.468	.330	
2B	1.354	.894	.733	.553	.453	.332	
2C	1.328	.854	.718	.521	.430	.304	
2D	1.327	.879	.714	.531	.444	.288	
2E	1.342	.883	.744	.556	.455	.328	
2F	1.298	.887	.728	.529	.449	.339	
x	1.336	.885	.733	.543	.449	.320	
s	.024	.020	.018	.018	.013	.020	
C.V.	1.82	2.25	2.51	3.37	2.86	6.17	3.16
3A	1.320	.969	.810	.605	.493	.376	
3B	1.336	.969	.778	.627	.510	.372	
3C	1.358	.946	.805	.622	.516	.376	
3D	1.319	.970	.776	.618	.497	.376	
3E	1.308	.951	.801	.625	.493	.374	
3F	1.282	.919	.821	.632	.520	.373	
x	1.320	.954	.798	.621	.505	.374	
s	.026	.020	.018	.009	.012	.002	
C.V.	1.94	2.10	2.25	1.51	2.38	0.47	1.78
4A	1.406	1.001	.841	.658	.519	.381	
4B	1.312	.970	.835	.648	.548	.397	
4C	1.368	.982	.816	.634	.521	.406	
4D	1.370	.980	.837	.655	.532	.413	
4E	1.389	.988	.828	.641	.518	.380	
4F	1.329	1.010	.848	.661	.518	.378	
x	1.362	.988	.834	.649	.526	.392	
s	.036	.015	.011	.010	.012	.015	
C.V.	2.62	1.48	1.33	1.61	2.28	3.82	2.19
x Singles	1.292	.919	.774	.589	.495	.369	
S.D. Singles	.090	.067	.057	.052	.030	.032	
C.V. Singles	6.97	7.34	7.32	8.77	6.14	8.67	3.03

Table 5. Comparison of Sulfamethazine Concentrations in Pig Blood by TLC and ELISA (ppm).

Date	Day	Pig 127-12 TLC	Pig 127-12 ELISA	Pig 127-13 TLC	Pig 127-13 ELISA	Pig 217-13 TLC	Pig 217-13 ELISA	Pig 217-10 TLC	Pig 217-10 ELISA
April 25	0	0	0	0	0	0	0	0	0
April 27	2	7.9	10.95	7.4	11.7	7.0	8.7	5.9	9.6
April 30	5	6.3	9.3	7.5	11.4	7.2	14.1	6.8	13.05
May 4	9	6.7	8.25	9.4	11.55	8.8	8.25	8.0	9.15
May 6	12	8.2	ND[1]	9.4	ND	17.1	ND	10.2	ND
May 9	14	8.4	5.6	9.6	6.8	10.6	11.2	8.0	8.4
May 11	16	0.56	1.55					1.6	1.7
May 14	19							0.09	0.08
May 18	23							0.01	0.08
May 21	26							0.01	0.04
May 25	30							0.00	0.025

Correlation Coefficient = 0.894
Y - Intercept = 0.539
Slope = 1.154

[a] ND - Not Determined

Four pigs were fed swine feed containing 110 ppm sulfamethazine for 14 days. Bleedings were made on different days and samples were sent to the Western Laboratory US Department of Agriculture Food Safety Inspection Service (Alameda, CA) (USDA-FSIS) for analysis by TLC (29). Identical samples were retained and tested at Neogen Corporation (Lansing, MI) using a quantitative ELISA and serum controls containing 0, 50, 100, 200, 400 and 500 ppb sulfamethazine. (Table 5)

Similar results were obtained using TLC and ELISA for analysis of the blood samples for sulfamethazine concentrations. Both methods required extensive sample dilution (>1/150) prior to analysis. This may account for differences in results between the methods. Variations from day to day, with slight decreases in values with subsequent increases in values on the following testing can be seen with values obtained by both methods. A correlation of 89.4% was obtained with a slope of 1.154 and y-intercept of 0.539.

The information in Table 6 shows that the mean recovery of sulfamethazine out of spiked serum is 100.5% with a standard deviation of 9.2 over 24 samples. This gives a coefficient of variation around these samples of 9.2%. This data indicates that measurement of sulfamethazine in swine serum using the ELISA is a very effective technique from a recovery standpoint. The issue of the recovery of sulfamethazine from whole blood is only relevant to the ability to recover serum from the blood since the test media will be sera and the information to support the use of this kind of sample to evaluate carcasses has all been performed in serum.

The ability to recover serum from the heart of an animal after slaughter is an important consideration from the standpoint that the blood could become very difficult to process after sitting for long periods of time in vivo. Therefore, a study was performed to look at long lengths of time to see if even the very longest time frames would cause problems.

Table 6. Results Obtained Using a Quantitative Format of the ELISA.

Assay	Conc. Spiked at (ppb)	Percent O.D. of Absorbance Sample	(B/Bo)	Results Read As (ppb)	Percent Recovery	Standard Curve (O.D. 650 nm) Concentration	
1	100	.651	53	118	118	0	1.224
	75	.740	60	84	112	50	0.835
	100	.666	54	112	112	100	0.687
	125	.623	51	130	104	200	0.513
	250	.442	36	270	108	300	0.425
	300	.411	34	300	100	500	0.305
	350	.378	31	355	101		
	300	.419	34	300	100		
	100	.677	55	110	110		
	75	.745	61	82	109		
	100	.661	54	115	115		
	125	.664	54	115	92		
	250	.474	39	235	94		
	300	.446	36	275	92		
	350	.426	35	290	83		
	300	.457	37	260	87		
2	100	.623	58	100	100	0	1.074
	75	.690	64	71	95	50	0.755
	100	.612	57	105	105	100	0.612
	125	.586	55	125	100	200	0.467
	250	.432	40	245	98	300	0.401
	300	.405	38	270	90	500	0.300
	350	.379	35	320	91		
	300	.401	37	290	97		

Mean (n=24)   100.5%
Standard Deviation   9.2%
Coefficient Variation   9.2%

There was no indication that time frames up to 4 hours created a problem for the recovery of serum out of the heart of slaughtered animals and, therefore, all shorter times are obviously not a problem.

The ability of the test to detect positive samples as positive was determined quantitatively and qualitatively using spiked samples near the decision points. In order to correct for a high percentage of false negative readings near the decision points (100 ppb low standard and 400 ppb high standard) both the low standard and high standards were adjusted downwards by 25% to 75 and 300 ppb sulfamethazine, respectively. A concentration of sulfamethazine greater than or equal to 100 ppb in the serum means that the liver contains violative levels of the drug ($\geq$100 ppb), while a concentration of SMZ of $\geq$400 ppb in the serum means that the muscle tissue contains violative levels of SMZ $\geq$100 ppb. And adjustment in the low and high standards would make the test err on the false positive side which is the desired outcome. False negative readings would allow contaminated organs and carcasses to reach the market place. A false positive reading would require retesting and confirmation by another method.

Table 7.   Verification of Adjusted Controls.

| | | | | Absorbance at 650 nm | | | | |
Run	LS[a]	0	100	200	300	400	500	HS[b]
1	.736	1.149	.702	.491	.403	.375	.291	.391
2	.585	.940	.470	.373	.323	.283	.269	.374
3	.701	.924	.561	.441	.353	.356	.233	.340
4	.732	1.108	.612	.481	.399	.327	.324	.366
5	.597	1.157	.607	.431	.362	.307	.262	.380
6	.575	1.010	.537	.435	.374	.293	.245	.321
7	.646	1.084	.586	.443	.361	.300	.274	.353
8	.713	1.168	.576	.432	.349	.287	.235	.339
9	.686	1.189	.589	.417	.335	.277	.240	.341
10	.526	1.014	.466	.332	.317	.246	.226	.292
11	.626	1.128	.561	.421	.362	.307	.273	.364
12	.682	1.150	.621	.440	.350	.281	.243	.358
x	.648	1.085	.574	.428	.356	.303	.260	.352
SD	.068	.087	.062	.041	.025	.034	.027	.026
CV	10.5	8.0	10.8	9.6	7.0	11.2	10.4	7.4

[a]   LS low standard = 75 ppb
[b]   HS high standard = 300 ppb

Table 7 shows the absorbance values obtained quantitatively with an overall coefficient of variation (CV) of 9.4%. There is excellent assay repeatability. Tables 8 and 9 show the qualitative results of the data presented in Table 7. Table 8 shows the sensitivity when the AgriScan is blanked on the low standard (98.3%) and Table 9 shows the high degree of sensitivity when the machine is blanked on the high standard (91.7%). Tables 10 and 11 present data which were generated when the test was run with spiked samples containing SMZ levels near the low and for high standard concentrations.

For Studies I and III the sensitivity again was high when the machine was blanked against both the low standard (93.8%, 100%) and the high standard (100%, 100%). For Study II the sensitivity was 100% when results were read against the low standard, there were no results obtainable for Study II when compared to the high standard. All of the spiked sample concentrations were less than 300 ppb and were seen as negative.

The ability of the test to read negative samples correction was also determined (Tables 8-10). There was a high degree of specificity when results were read against the low standard (Table 8 and 10). The test specificity obtained was 100%, regardless of which study was performed.

The specificity was not quite as high when results were obtained. It varied depending on what the concentration of the spiked samples were. The specificity was high in Table 9 (95.8%) and in Studies I and II presented in Table 11 (95% and 100%). In Study III samples were spiked at concentrations at or near the concentration of SMZ in the high standard. The specificity was 66.7%. There were two false positives at 300 ppb and two false negatives at 350 ppb. However, at levels of ≥400 ppb SMZ all results were positive. There were no false negatives. The false negatives at 350 ppb are really not a problem, since the sample really still is negative, since the tissue would not contain violative SMZ levels. The false positives at 300 ppb would lead to additional testing, which is preferred to erring on the false negative side.

Table 8. Qualitative Screening of Spiked Samples - Machine Blanked Against Low Standard[a].

0	100	200	300	400	500
-5	+2	+5	+5	+5	+5
-5	+3	+5	+5	+5	+5
-5	+3	+5	+5	+5	+5
-5	+4	+5	+5	+5	+5
-5	-1[b]	+4	+5	+5	+5
-5	+1	+4	+5	+5	+5
-5	+2	+5	+5	+5	+5
-5	+4	+5	+5	+5	+5
-5	+3	+5	+5	+5	+5
-5	+2	+5	+5	+5	+5
-5	+2	+5	+5	+5	+5
-5	+2	+5	+5	+5	+5

Sensitivity   11/12 12/12 12/12 12/12 12/12 = 59/60 = 98.3%
Specificity 12/12 ---   ---   ---   ---   ---   = 12/12 = 100%
              Overall Accuracy = 71/72 = 98.6%

[a] Low Standard = 75 ppb
[b] False Negative
0, -1 to -5, negative reading
+1 to +5, positive reading

Table 9. Qualitative Screening of Spiked Samples - Machine Blanked Against High Standard.

0	100	200	300	400	500
-5	-5	-2	-1	+1	+2
-5	-2	0	+1[b]	+3	+3
-5	-5	-3	-1	-1[c]	+2
-5	-4	-3	-1	+1	+1
-5	-4	-1	+1[b]	+2	+3
-5	-4	-3	-2	0[c]	+2
-5	-4	-2	-1	+1	+2
-5	-4	-3	-1	+1	+3
-5	-4	-1	0	+2	+3
-5	-3	-1	-1	+1	+2
-5	-4	-1	0	+1	+2
-5	-4	-2	0	+1	+2

Sensitivity ---   ---   ---   ---   10/12 12/12 = 22/24 = 91.7%
Specificity 12/12 12/12 12/12 10/12 ---   ---   = 46/48 = 95.8%
              Overall Accuracy = 68/72 = 94.4%

[a] High Standard =.300 ppb
[b] False Positives
[c] False Negatives
0, -1 to -5, negative reading
+1 to +5, positive reading

Table 10. Qualitative Analysis of Spiked Samples - Machine
Read on Low Standard[a].

___

Study I

	0	50	100	200	300	400	
	−5	−3	0	+5	+5	+5	
	−5	−3	+1	+5	+5	+5	
	−5	−1	+4	+5	+5	+5	
	−5	−3	+3	+5	+5	+5	
Sensitivity	---	---	3/4	4/4	4/4	4/4 =	15/16 = 93.8%
Specificity		4/4	4/4	---	---	---	--- = 8/8 = 100%

Overall Accuracy = 23/24 = 95.8%

Study II

	0	25	50	100	150	200	
	−5	−5	−3	+2	+3	+5	
	−5	−5	−4	+5	+5	+5	
	−5	−5	−2	+4	+4	+5	
	−5	−5	−2	+3	+5	+5	
Sensitivity	---	---	---	4/4	4/4	4/4 =	12/12 = 100%
Specificity	4/4	4/4	4/4	---	---	--- =	12/12 = 100%

Overall Accracy = 24/24 = 100%

Study III

	0	300	350	400	450	500	
	−5	+5	+5	+5	+5	+5	
	−5	+5	+5	+5	+5	+5	
	−5	+5	+5	+5	+5	+5	
	−5	+5	+5	+5	+5	+5	
Sensitivity	---	4/4	4/4	4/4	4/4	4/4 =	20/20 = 100%
Specificity		4/4	---	---	---	---	--- = 4/4 = 100%

Overall Accuracy = 24/24 = 100%

___

[a] Low Standard = 75 ppb
0, −1 to −5, negative reading
+1 to +5, positive reading

Table 11. Qualitative Analysis of Spiked Samples - Machine
Read on High Standard.

Study I

	0	50	100	200	300	400
	-5	-5	-4	-1	0	+2
	-5	-5	-4	-2	0	+2
	-5	-5	-4	-1	+1	+3
	-5	-5	-3	-1	0	+3
Sensitivity	---	---	---	---	---	4/4 = 4/4 = 100%
Specificity	4/4	4/4	4/4	4/4	3/4	--- = 19/20 = 95%

Overall Accuracy = 23/24 = 95.8%

Study II

	0	25	50	100	150	200
	-5	-5	-5	-4	-3	-2
	-5	-5	-5	-4	-3	-1
	-5	-5	-5	-3	-3	0
	-5	-5	-5	-4	-4	-2
Sensitivity	---	---	---	---	---	---
Specificity	4/4	4/4	4/4	4/4	4/4	4/4 = 24/24 = 100%

Overall Accuracy = 24/24 = 100%

Study III

	0	300	350	400	450	500
	-5	+1	0	+2	+2	+2
	-5	+1	+2	+3	+3	+4
	-5	0	+1	+2	+2	+2
	-5	0	0	+2	+2	+2
Sensitivity	---	---	---	4/4	4/4	4/4 = 12/12 = 100%
Specificity	4/4	2/4	2/4	---	---	--- = 8/12 = 66.7%

Overall Accuracy = 20/24 = 83.3%

---

[a] High Standard = 300 ppb
0, -1 to -5, negative reading
+1 to +5, positive reading

Table 12. Monoclonal Antibody Sensitivity for Ivermectin[a]

Monoclonal Antibody[b]	Ivermectin I_{50}[c], ppb	Slope
	3-14	0.9-1.0
	6-20	1.2-1.7
	8-10	0.9-1.2
	10-31	1.0-1.2
	12-15	1.3-1.4

[a]The data represent the range of dose responses during 3 months of competitive ELISA experiments.
[b]All antibodies are subclass IgG, K.
[c]I_{50} half maximal inhibition.

It should be noted that a temperature study was performed at 60, 65, 75, 80 and 90°F to simulate possible temperatures in slaughter houses. The test performed well over the wide range of temperatures (data not shown).

Overall accuracy of the test was determined to be 98.6% (142/144) when all samples were tested against the low standard. The overall accuracy of the test was 93.7% (135/144) when the samples were tested against high standard. This means that in over 90% of the samples tested correct responses will be obtained and will decrease chances for carcasses containing violative SMZ levels to reach the market place. This accuracy is excellent for a screening format.

Avermectins. Avermectins are naturally occurring antibiotics secreted by the actinomycete Streptomyces avermetilis (34). They are active against helminths and arthorpods, but did not exhibit significant antibacterial or antifungal activity (35).

Avermectins, namely ivermectin, are being used regularly for livestock pest management, veterinary medicine, parasitology.

Discussion of this ELISA was included as an example since it has a detection limit similar to the classical chemical methods (16). However, the classical extraction and cleanup methods for ivermectin require 41 steps, while the method described in this manuscript only requires nine steps (16).

Monoclonal antibodies were produced to ivermectin. The antibodies recognize both ivermectin (and abamectin, the major avermectin used in agricultural applications).

A description is provided of how microtiter plates were selected, the plate coating conditions tried, the block agents tested, incubation times and temperatures tried, and testing of enzyme conjugated antibodies to optimize assay conditions. Monoclonal antibody C4D6 (Table 12) bound ivermectin in buffers containing up to 30% (v/v) or organic solvent. The antibody C4D6.1 was able to bind ivermectin in PBS-Tween contain 5, 10, or 20% (v/v) acetonitrile, acidified acetonitrile or dimethylformamide with no significant differences (16). Monoclonal antibody C4D6 bound to ivermectin in PBS-Tween containing up to 30% methanol or acetonitrile with out major differences in the ELISA performance or 50% point of inhibition of antibody binding. Dose response curves which were nearly identical when the test was carried out with PBS-Tween with 5% dimethylsulfoxide or, 10% tetrahydrofuran, 5% acetonitrile and 5% methanol. Use of 20% dimethyl sulfoxide decreased the ELISA rate a little but did not change the slope, while use of more than 5% tetrahydrofuran caused increased scatter at doses <1.0 ppb ivermectin (16). The ability to use this variety of solvents facilitates the development of immunoassay for agricultural residues recovered form liquid or solid phase. The limit of detection for the ivermectin standards in the ELISA was about 0.5 ppb with half maximal inhibition (I_{50}) being about 3 ppb.

These antibodies appear to be suitable for detecting and quantifying avermectin residues in veterinary samples.

ACKNOWLEDGEMENTS

The author wishes to acknowledge the following individuals for their able technical and clerical assistance.

1. The group effort by the following individuals to develop and validate the ELISA for detection of sulfamethazine in swine blood (serum).

   a. Drs. Brinton Miller and James Harness as well as Mrs. Mary Heine and Ms. Christine Berns of Neogen for their contributions in test development.

   b. Dr. Elwyn Miller, Michigan State University, Department of Animal Science, for help in carrying out the swine dosing study.

   c. Ms. Estrella Agamata, USDA/FSIS, Western Regional Laboratory for performing the TLC on the serum samples to obtain a correlation between the methods.

2. Dr. Stanley Katz, Rutgers University, Department of Biochemistry and Microbiology and Dr. Vipin Agarwal, The Connecticut Experimental Station for encouraging the authors' participation in the symposium, and;

3. Mrs. Lisa Summers for preparation of this manuscript.

REFERENCES

1. Newsweek, Warning: Your food, nutritious and delicious may be hazardous to your health, March 29, 1989.
2. Wall Street Journal, Milk is found tainted with range of drugs farmers give cattle, Friday, Dec. 29, 1989.
3. National Hog Farmer, Sulfamethazine: A Spoon of Trouble, April 15, 1989.
4. L.F. Knudsen and W.A. Randall, Penicillin assay and its control chart analysis, J. Bacteriol., 50:187 (1945).
5. F. Kavanaugh, "Analytical Microbiology", Vol. II, Academic Press, New York (1972).
6. A.C. Bratton, E.K. Marshall, Jr., D. Babbitt and A.R. Hendrickson, A new coupling component for sulfanilamide determination, J. Biol. Chem., 28:537 (1939).
7. E.K. Marshall, Jr. and D. Babbitt. Determination of sulfanilamide in blood and urine, J. Biol. Chem., 22:263 (1937-1938).
8. M.F. Delaney, The "chromatographic uncertainty principle", Liquid Chromatogr., 2:85 (1984).
9. T.G. Alexander, Gas Chromatographic Analysis, in "Modern Analysis of Antibiotics", A. Aszalos, ed., Marcel Dekker, New York (1986).

10. Z.M. Dinya and F.J. Sztaricskai, Ultraviolet and light absorption spectrometry, in "Modern Analysis of Antibiotics, A. Aszalos, ed, Marcel Dekker, New York (1986).

11. R.G. Smith, Mass Spectrometric Analysis, in "Modern Analysis of Antibiotics", A. Aszalos, ed., Marcel Dekker, New York (1986).

12. D.E. Dixon, S.J. Steiner and S.E. Katz, Immunological Approaches, in: "Modern Analysis of Antibiotics", A. Aszalos, ed, Marcel Dekker, New York (1986).

13. J.W. Goding, Antibody production by hybridomas, J. Immunol. Meth., 39:285 (1980).

14. G.S. Campbell, R.P. Mageau, B. Schwab and R.W. Johnston, Detection and quantitation of chloramphenicol by competitive enzyme-linked immunoassay, Antimicrob. Agents and Chemo., 25:205 (1984).

15. A.H.M. Schotman, Mechanism of the reaction of carbodiimides iwth carboxylic acids. Receil des Travax Chim des Pays. Bas, 110:319 (1991).

16. D.J. Schmidt, C.E. Clarkson, T.A. Swanson, M.L. Egger, R.E. Carlson, J.E. VanEmon and A.E. Karu, Monoclonal Antibodies for Immunoassay of Avermectins, J. Agric. Food Chem., 38:1763 (1990).

17. D.B. Berkowitz and P.W. Webert, Enzyme immunoassay based survey of prevalence of gentamicin in serum of marketed animals, J. Assoc. Off. Anal. Chem., 69:437 (1986).

18. D.E. Dixon-Holland and S.E. Katz, Use of a competitive direct enzyme-linked immunosorbent assay on detection of sulfamethazine residues in swine tissue and urine, J. Assoc. Off. Anal. Chem., 71:1137 (1989).

19. D.E. Dixon-Holland S.E. Katz, Direct competitive enzyme-linked immunosorbent assay for sulfamethazine residues in milk, J. Assoc. Off. Anal. Chem., 72:447 (1989).

20. D.E. Dixon-Holland and S.E. Katz, A competitive direct enzyme-linked immunosorbent screening assay for the detection of sulfonamide contamination of animal feeds, J. Assoc. Off. Anal. Chem., 72:784 (1991).

21. J.R. Fleeker and L.J. Lovett, Enzyme immunoassay for screening sulfamethazine residues in swine blood, J. Assoc. Off. Anal. Chem., 68:172 (1985).

22. B.F. Erlanger, Principles and methods for the preparation of drug protein conjugates for immunological studies, Pharmacol. Reviews, 25:271 (1973).

23. J.J. O'Rangers, Development of Drug Residue Immunoassays, in: "Immunochemical Methods for Environmental Analysis", J.M. Van Emon and R.O. Mumma, eds., American Chemical Society, Washington, D.C. (1990).

24. B. Albini and A. Andres, Immunoelectronmicroscopy, in: "Principles of Immunology", N.R. Rose, F. Milgrom and C. VanOss, eds., Second edition, MacMieran Publishing.

25. P. Rohner, M. Schallibaum and J. Nicolet, Detection of penicillin G and its benzyl penicilloyl (BPO) - derivatives in cow milk and serum by means of an ELISA, J. Food Prot., 48:59 (1985).

26. Milk and Dairy Beef Quality Assurance Program, American Veterinary Medical Association and National Milk Producers Federation, Agri Education, Inc. (1991).

27. J.F.M. Nouws, F. Reek, M.M.L. Aerts, M. Baakman and J. Laurensen, Monitoring slaughtered animals for chloramphenicol residues by an immunoassay test kit (Quik-card®), Archiv fur Lebensmittelhygiene, 38:1 (1987).

28. D.E. Dixon-Holland, M.A. Heine, E.E. Smith and B.M. Miller, A rapid screening ELISA for detection of sulfamethazine residue in swine serum, Presented at the 104th Annual Meeting of the Association of Official Analytical Chemists (1990).

29. M.H. Thomas, K.E. Soroka and S.H. Thomas, _J. Assoc. Off. Anal. Chem._,
    66:88 (1986).
30. R.B. Ashworht, R.L. Epstein, M.H. Thomas and L.T. Frobish,
    Sulfamethazine blood/tissue correlation study in swine, _Am. J._
    Vet. Res., 47:2596 (1985).
31. V.W. Randecker, J.A. Reagan, R.E. Engel, D.L. Soderberg and J.E.
    McNeal, Serum urine as predictors of sulfamethazine levels in
    swine muscle, liver and kidney, _J. Food Prot._, 50:115 (1987).
32. D.D. VanHoweling and F.J. Kingma, The use of drugs in animals raised
    for food, _Am. J. Vet. Res._, 155:2197 (1969).
33. W. Horwitz, Review of analytical methods for sulfonamides in foods and
    feeds, II. Performance characteristics of sulfonamide methods,
    J. Assoc. Off. Anal. Chem., 64:814 (1981).
34. J.C. Chabala, H. Mrozik, R.L. Tolman, P. Eskola, A. Lusi, L.H.
    Peterson, M.F. Woods and M.H. Fisher, Ivermectin - a new broad-
    spectrum antiparastic agent, _J. Med. Chem._, 23:1134 (1980).
35. R.W. Burg, B.M. Miller, E.E. Baker, J. Birnbaum, S.A. Currie, R.
    Hartman, Y. Kong, R.L. Monghan, G. Olson, I. Putter, J.B. Tunac,
    H. Wallick, E.O. Stapley, R. Oiwa, and S. Omura, Avermectins, a
    new family of potent anthelminthic agents: producing organism
    and fermentation. _Antimicrob. Agents Chemther._ , 15:361 (1979).

EVALUATION AND TESTING OF CHARM TEST II RECEPTOR ASSAYS FOR THE DETECTION OF ANTIMICROBIAL RESIDUES IN MEAT

Gary O. Korsrud, Craig D. C. Salisbury, Adrian C. E. Fesser and James D. MacNeil

Health of Animals Laboratory, Agriculture Canada, 116 Veterinary Road, Saskatoon, Saskatchewan, S7N 2R3

ABSTRACT

Charm Test II receptor assays for β-lactams, sulfonamides, streptomycin and erythromycin were tested with standard solutions of targeted antimicrobials and with fortified kidney, liver and muscle samples. The degree of response obtained with the β-lactam assay varied among equimolar amounts of 8 β-lactams. The sulfa assay was positive, with varying response, to equimolar amounts of four different sulfonamides and negative for trimethoprim. The streptomycin assay also detected dihydrostreptomycin, but was negative for gentamicin and neomycin. The erythromycin assay was negative for oleandomycin but positive for tilmicosin and tylosin. No evidence of cross reactivity was detected among the compounds targeted by the four assays.

These assays were then used to analyze samples of kidney from 264 cull Holstein cows that were slaughtered at a local packing plant. Liver and diaphragm muscle tissue from carcasses having positive kidney results were also tested. Kidneys from 12 cows were positive for streptomycin by the Charm Test II. Four livers and one diaphragm muscle from the same 12 cows were also positive. The kidneys and livers from four cows and the diaphragm muscle from three cows were positive by the Charm Test II for β-lactams. Penicillin G was confirmed by high performance liquid chromatography (HPLC) in three of these kidney-liver pairs and in two of the associated diaphragm muscles. Tissues from one cow tested positive for sulfonamides, with sulfamethazine confirmed in liver, kidney and diaphragm muscle by thin layer chromatography. All kidneys tested negative using the erythromycin test.

INTRODUCTION

The Charm Test II assays were first developed to test for various antimicrobials in milk (1 - 3). Recently Charm Sciences Inc. has modified them for use in screening tissues for β-lactams, sulfonamides, streptomycin, erythromycin and chloramphenicol. The assays are based on a binding reaction between drug functional groups and receptor sites on microbial cells. Radio-labelled target drug is added and competes with any incurred drug residue present in the sample to bind at the receptor sites. A liquid scintillation counter is used to measure bound ^{14}C or ^{3}H from the labelled drug. The greater the amount of drug present in the sample, the lower the counts, as any labelled drug not bound to receptor sites is removed by rinsing of the substrate prior to counting.

Van Dresser and Wilcke (4), Franco et al. (5), and Guest and Paige (6) have reported that cull dairy cows had the highest incidence of antibiotic residues in studies they conducted. We therefore applied these tests to tissues from cull Holstein dairy cows to allow a practical evaluation of the kits, and to establish the incidence of drug residues in this population as detected with this technology.

Analysis of Antibiotic Drug Residues in Food Products of Animal Origin
Edited by V.K. Agarwal, Plenum Press, New York, 1992

75

EXPERIMENTAL

Charm Test II tissue assay kits for β-lactams, sulfonamides, streptomycin and erythromycin were obtained from Charm Sciences Inc., Malden, MA, and used according to manufacturer's protocols. Available laboratory centrifuges and a Beckman 5801 liquid scintillation counter were substituted for the centrifuge and liquid scintillation counter normally supplied by the test kit manufacturer. Graded levels of each antimicrobial were added to blank kidney, liver and muscle samples. Detection limits were based on the fortification level required to produce a 30% decrease in counts for the fortified sample compared to the blank tissue for sulfamethazine, streptomycin and erythromycin, and a decrease of 20% for penicillin G, based on triplicate determinations. We considered such decreases sufficient to consider samples positive. Tests for cross reactivity were conducted in triplicate with fortified tissues using graded levels of each antimicrobial in excess of the levels used for the usual standards. Within-group reactivity was determined in the same fashion.

Kidney, liver and diaphragm muscle samples were collected from 264 cull Holstein cows slaughtered at a local federally inspected packing plant between January 22 and March 26, 1991. All cull Holstein cows slaughtered on a given collection day were sampled. Samples were stored at -20°C prior to testing at the laboratory. Kidney samples were tested with the Charm Test II receptor assays for β-lactams, sulfonamides, streptomycin and erythromycin. A single test was conducted on each sample with each assay kit. The four different tests were conducted on aliquots of a single extract from each sample. If the kidney sample was negative no further tests were done. Kidney samples that were positive were retested in duplicate and duplicate tests were conducted on liver and diaphragm muscle samples from the same cows. If the duplicate retest was negative the sample was considered negative.

Samples found positive for β-lactams using the Charm Test II assay were tested for penicillin G using a high performance liquid chromatographic method sensitive to 5 ppb levels in tissue (7), while samples which tested positive for sulfonamides were also analyzed by thin-layer chromatography with fluorescence detection (8). Samples which responded positively to the Charm Test II streptomycin method could not be confirmed, due to the lack of a suitable chemical determinative method.

RESULTS AND DISCUSSION

The detection limits obtained with penicillin G, sulfamethazine, streptomycin and erythromycin, respectively, were 5, 15, 100 and 150 parts per billion (ppb).

With the β-lactam assay the relative decreases in disintegrations per minute observed for equimolar amounts of potassium penicillin G, sodium penicillin G, procaine penicillin G, penicillin V, amoxicillin, ampicillin, cloxacillin and ceftiofur, respectively, were 100, 102, 102, 86, 82, 80, 35 and 77. The concentration of potassium penicillin G was equivalent to a 15 ppb spike. With the sulfa assay the relative decreases in disintegrations per minute observed for equimolar amounts of sulfamethazine, sulfadimethoxine, sulfathiozole and sulfadoxine, respectively, were 100, 240, 310 and 210. The concentration of sulfamethazine used was equivalent to a 50 ppb spike. Trimethoprim at 50 ppb did not interfere with the sulfa assay. The streptomycin assay also gave a positive response to dihydrostreptomycin, but was negative for 1,000 ppb of either gentamicin or neomycin. The erythromycin assay was negative for oleandomycin at 1,000 ppb. With the erythromycin assay the relative decreases in disintegrations per minute observed for erythromycin, tilmicosin and tylosin, respectively, were 100, 82 and 30. The concentrations used were equivalent to 300 ppb spikes. No evidence of cross reactivity was detected among the compounds targeted by the four assays. Concentrations of 2,000 ppm of streptomycin plus 400 ppm of sulfamethazine did not affect the β-lactam results. Similarly 2,000 ppm streptomycin or 400 ppm penicillin G did not affect the sulfa results. Concentrations of 400 ppb of penicillin G and sulfamethazine did not affect the streptomycin results and 300 ppb of penicillin G, sulfamethazine and streptomycin did not affect the erythromycin results.

Application of these assays to the kidneys collected from 264 cull Holstein cows slaughtered at a local packing plant yielded 12 cows (4.5%) that tested positive for streptomycin (Table 1). Five of those cows numbered 64 - 68 came from the same producer. Four of the livers and one diaphragm muscle from the same 12 cows were also positive for streptomycin using the Charm Test II assay.

Kidneys and livers from four cows (1.5%) were positive by the Charm Test II for β-lactams (Table 2). Penicillin G was confirmed by high performance liquid chromatography (HPLC) in three of those kidneys and livers. Diaphragm muscle samples from three of the same four cows were positive by the Charm Test II and two of these results were confirmed by HPLC analysis. There was, however, no correlation between the concentration of penicillin G in the tissue, as determined by liquid chromatographic assay, and the reduction in the counts observed for the Charm Test II assay result (Table 2). The concentrations of penicillin G detected, as determined by liquid chromatography, ranged from the detection limit of 5 ppb in a diaphragm muscle sample to 422 ppb in a liver sample. All tissues collected from one cow were positive for β-lactams by the Charm Test II assay, but contained no detectable penicillin G by liquid chromatography, suggesting the possibility that this animal may have contained residues of another β-lactam drug and needs further confirmation as the Charm Test II detects the entire β-lactam family. Since three tissues from the same animal are positive at quite high levels there is most likely some β-lactam present other than penicillin G. Alternatively, false positive results may have been obtained for the tissues for that particular animal.

The kidney and liver from one cow were positive for sulfonamides by the Charm Test II (Table 3). Sulfamethazine was confirmed in liver, kidney and diaphragm muscle from that cow, using thin layer chromatography, at levels ranging from the detection limit of 20 ppb in the diaphragm muscle to 80 ppb in the liver.

The Charm Test II for erythromycin was negative for all kidneys tested.

There were five streptomycin, one sulfonamide, two β-lactam and two erythromycin results that were positive for the first single determination but were negative when retested in duplicate. The decrease in counts was more than 40% for only two of those 10 kidneys. None of the diaphragm muscle or liver samples from those same 10 cows tested positive.

Table 1. Cull Holstein Cows Charm Test II for Streptomycin,
% of Control, a Total of 264 Cows Tested

Cow Number	Tissue		
	Kidney	Liver	Diaphragm muscle
26	19[a]	27[a]	88
49	55[a]	84	89
59	14[a]	18[a]	67[a]
64	34[a]	77	93
65	22[a]	42[a]	90
66	53[a]	83	83
67	18[a]	29[a]	81
68	39[a]	79	80
140	34[a]	75	104
145	54[a]	73	111
172	36[a]	94	111
218	65[a]	102	90
Number positive[a]	12	4	1

[a] A value of 70% or less of control is considered positive.

Table 2. Cull Holstein Cows Charm Test II, for β-lactams and High Performance Liquid Chromatography for Penicillin G, a Total of 264 Cows Tested

Cow Number	Charm Test II (% of Control)			High Performance Liquid Chromatography (ppb)		
	Kidney	Liver	Diaphragm muscle	Kidney	Liver	Diaphragm muscle
59	45[a]	20[a]	105	28	35	<5
88	31[a]	41[a]	74[a]	40	103	5.3
130	38[a]	39[a]	61[a]	<5	<5	<5
228	20[a]	17[a]	78[a]	257	422	19.5
Number Positive[a]	4	4	3	3	3	2

[a] A value of 80% or less of control for the Charm Test II is considered positive.

Table 3. Cull Holstein Cows Charm Test II and Thin Layer Chromatography for Sulfonamides, a Total of 264 Cows Tested

Cow Number	Charm Test II (% of Control)			Thin Layer Chromatography (ppb)		
	Kidney	Liver	Diaphragm muscle	Kidney	Liver	Diaphragm muscle
157	32[a]	44[a]	80	40[b]	80[b]	20[b]
Number Positive[a]	1	1		1	1	1

[a] A value of 70% or less of control for the Charm Test II is considered positive.

[b] Sulfamethazine

The total number of Charm Test II positive samples was comprised of 16 kidneys, 8 livers and 4 diaphragm muscles, representing 6.1%, 3.0% and 1.5%, respectively, of the animals tested. One cow tested positive for both streptomycin and β-lactams, the latter confirmed as penicillin G by liquid chromatography. Tritschler et al (9) reported 3% positive kidneys from 281 randomly selected cull dairy cows when the Swab Test on Premises, or STOP (10), was used for testing. The STOP is commonly used at packing plants to test slaughter carcasses for microbial inhibitors. Current Canadian tolerances for penicillin G and sulfonamides in cattle tissues are 50 ppb and 100 ppb, respectively, while no streptomycin or erythromycin residues are permitted. Based on our findings, tested animals in violation of these tolerances included one carcass for streptomycin residues, plus a total of 13 kidneys and 6 livers for streptomycin and/or penicillin G residues which would be subject to condemnation.

The Charm Test II detected positive samples in this field trial, showing a degree of specificity for antimicrobial groups of related compounds, combined with good sensitivity. No information is available from this study on the number of false negatives because the negative samples were not analyzed by the chemical methods. As no other screening tests were run comparatively, no comment can be made as to its performance relative to other screening tests, such as the STOP. It proved quick and easy to perform, requiring limited laboratory facilities. It does require a liquid scintillation counter. For drugs with established tolerance levels, it can serve only as a screening test because it is not quantitative and therefore then be supported by quantitative chemical methods. It is effective for drugs with a zero tolerance as only detection is required. Where chemical confirmatory tests were available for penicillin G and sulphonamides, 77% of the positive Charm Test II results were confirmed. All samples positive by the Charm Test II sulfa and β-lactam assays (excluding animal 130) were confirmed to contain sulfamethazine and penicillin G. Liver and kidney, known to contain antibiotics at levels higher than muscle, were strong positives with the Charm Test II. While diaphragm muscle tissue samples containing lower levels of drugs were negative or borderline positives with the Charm Test II. The Charm Test II for β-lactams detected penicillin G in one diaphragm muscle sample assayed by liquid chromatography at 5.3 ppb, with the Charm Test II assay result for this sample being 74% of the control value. The Charm Test II for sulfonamides did not detect sulfamethazine at 20 ppb in a diaphragm muscle, using our criteria for a positive, but did respond to levels of 40 ppb and 80 ppb in the kidney and liver, respectively, of that animal. These tests therefore appear capable of detecting residues in tissue below current Canadian tolerances for penicillin G and sulfonamides. However, where quantitative analysis is required to determine if a residue is above tolerance, the use of additional laboratory assays, such as the ones used in this study, is required.

ACKNOWLEDGEMENTS

The authors express their sincere thanks to B. Petford, A. Malcolm, W. Chan and T. Duff for technical assistance. The cooperation of Drs. G. Graham and A. Klemer and other meat inspection staff at the packing plant is gratefully acknowledged.

LITERATURE CITED

1. Suhren, G. and W. Heeschen, Detection of antibiotics in milk with a modified microbial receptor assay (Charm Test II). Milchwissenschaft 42(8) 493-496 (1987).
2. Collins-Thompson, D. L., D. S. Wood and I. Q. Thomson, Detection of antibiotic residues in consumer milk supplies in North America using the Charm Test II procedure. J. Food Protection 51(8) 632-633 and 650 (1988).
3. Carlsson, A. and L. Björck, Detection of antibiotic residues in herd and tanker milk. A study of the Charm Test II. Milchwissenschaft 44(1) 7-10 (1989).
4. Van Dresser, W. R. and J. R. Wilcke, Drug residues in food animals. J. Amer. Vet. Med. Assoc. 194(12) 1700-1710 (1989).
5. Franco, D. A., J. Webb and C. E. Taylor, Antibiotic and sulfonamide residues in meat: Implications for human health. J. Food Prot. 53(2) 178-185 (1990).
6. Guest, G. B. and J. C. Paige, The magnitude of the tissue residue problem with regard to consumer needs. J. Amer. Vet. Med. Assoc. 198(5) 805-808 (1991).
7. Boison, J.O., C.D.C. Salisbury, W. Chan and J.D. MacNeil, Determination of penicillin G residues in edible animal tissues by liquid chromatography. J. Assoc. Offic. Anal. Chem. 74(3) 497-501 (1991).
8. Thomas, M., K.E. Soroka and S.H. Thomas, Quantitative thin layer chromatographic multi-sulfonamide screening procedure. J. Assoc. Offic. Anal. Chem. 66(4) 881-883 (1983).
9. Tritschler, J.P., II, R.T. Duby, S.P. Oliver and R.W. Prange, Microbiological screening tests to detect antibiotic residues in cull dairy cows. J. Food Protect. 50(2) 97-102 (1987).
10. Johnston, R.W., R.H. Reamer, E.W. Harris, H.G. Fugate and B. Schwab, A new screening method for the detection of antibiotic residues in meat and poultry tissues. J. Food Protect. 44(11) 828-831 (1981).

IMMUNOCHEMICAL METHODS IN THE ANALYSIS OF VETERINARY DRUG RESIDUES

Nel Haagsma and Cornelis van de Water

Department of the Science of Food of Animal Origin
Faculty of Veterinary Medicine, University of Utrecht
P.O. Box 80.175, 3508 TD Utrecht, The Netherlands

The application of immunochemical methods in veterinary drug residues is discussed. These methods concern (i) immunochemical determinations such as ELISAs for rapid screening purposes and (ii) immunoaffinity clean-up techniques for crude samples or sample extracts prior to a physicochemical determination. Examples of these methods are given. Special attention is paid to the detection and/or determination of chloramphenicol residues by these techniques. This concerns a rapid detection method using a streptavidin-biotin ELISA and an immunoaffinity clean-up, both off-line and by means of a fully automated HPLC system.

INTRODUCTION

In the course of the last decade a large number of chemical methods has been developed for the analysis of veterinary drug residues in food. Physicochemical methods are used at present for screening, quantification and confirmation of residues resulting from many different types of drugs. In recent years, considerable progress has been made particularly with respect to rapid and easy sample preparation procedures using solid-phase extraction - both off- and on-line - preceding the final analysis[1].

Apart from this, immunochemical methods are gaining popularity in veterinary drug analysis. These methods concern (i) immunochemical determinations such as enzyme-linked immunosorbent assays (ELISAs) for rapid screening purposes, in which the presence of a certain drug can be established, and (ii) immunoaffinity clean-up for crude samples or crude sample extracts prior to a physicochemical determination, which makes very low detection limits possible.

All immunochemical methods are based on the highly specific binding between the analyte and the antibody raised against this compound. Veterinary drugs, being low-molecular weight compounds, generally do not give an immuno-

Analysis of Antibiotic Drug Residues in Food Products of Animal Origin
Edited by V.K. Agarwal, Plenum Press, New York, 1992

81

genic response in an animal. Such haptenes are defined as "chemical moieties which can react specifically with the appropriate antibody but are not immunogenic"[2]. To produce such a response, these compounds have to be coupled to a large molecule such as bovine serum albumine (BSA).

After immunization of an animal a conventional, or polyclonal, antiserum can be produced, containing antibodies of varying isotype, affinity and specificity that potentially recognize the different epitopes present in the antigen (hapten). A polyclonal antiserum can be regarded as a mixture of monoclonal antibodies, i.e. antibodies produced by the clone of a simple B lymphocyte.

In 1975, Köhler and Milstein[3] reported a technique for the production of monoclonal antibodies. In this 'hybridoma technology', immortal antibody-producing cells are obtained after fusion of tumour cells with spleen cells of an immunized mouse. After a period of growth the hybridoma cultures can be screened for specific antibody production. In this way, antibodies of the desired affinity and specificity can be selected. Moreover, monoclonal antibodies, in comparison with polyclonal antibodies, possess the advantage of a continuous supply of constant quality. On the other hand, producing monoclonal antibodies is much more expensive than the production of polyclonal antibodies. This may be the reason that polyclonal antibodies are used in the majority of applications to veterinary drug residue analysis. Moreover, in some cases polyclonal antibodies will give lower limits of detectability.

IMMUNOCHEMICAL ASSAYS

The first immunochemical methods, i.e. radioimmunoassays (RIAs) for the determination of hormonal substances in blood, were developed around 1960[4,5] From this development, there has been an enormous growth in the application of these techniques, particularly in the field of clinical chemistry. This was mainly due to the introduction of enzyme-labelled immunoassays by Engvall and Perlman[6] and Van Weemen and Schuurs[7].

In food analysis the introduction of immunochemical methods was comparatively slow[8], but in the course of the last decade the application of immunoassays in this field has increased considerably. This was also the case with the analysis of veterinary drug residues.

For these residues, RIAs were particularly applied when low levels had to be determined and where determination by physicochemical methods was difficult, e.g. diethylstilbestrol in urine[9], carazolol in blood and urine[10] and chloramphenicol in eggs[11]. At present, enzyme immunoassays have been applied for the detection and/or determination at the low μg kg^{-1} level. There are several advantages in using enzyme immunoassays such as ELISAs over RIAs; (i) the long-term stability of the enzyme label, (ii) the cheap and commonly available equipment, (iii) the absence of radiation hazards, and (iv) the feasibility under field conditions.

As for drugs, competitive ELISAs are used in particular. Within these assays, different modes of performance are possible. The analyte (hapten) can be labeled with an enzyme as well as the antibody (Fig. 1). The sandwich ELISA is not used generally for small molecules, as a result of steric hindrance. For macromolecules such as proteins, sandwich as well as competitive ELISAs are described.

ELISAs can be used for both qualitative and quantitative purposes. The specificity of these methods decreases the need for a thorough clean-up; therefore sample preparation procedures in immunoassays generally are rather simple, making ELISAs particularly suitable for the screening of large series of samples.

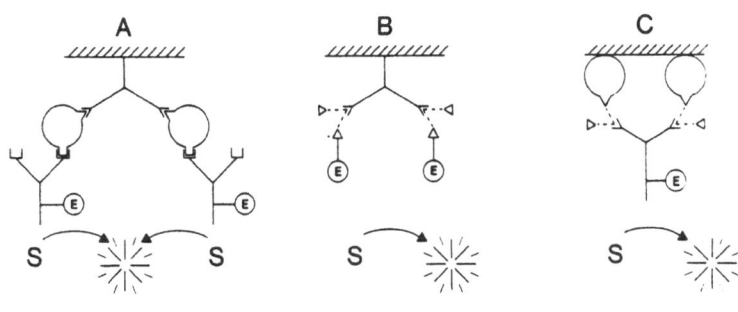

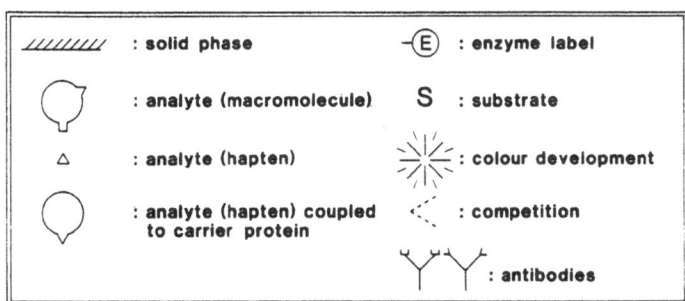

Fig. 1.   Schematic presentation of different ELISAs applied in food analysis. A: sandwich ELISA, B: competitive ELISA using enzyme-labeled analyte, C: competitive ELISA using enzyme-labeled antibody.

If extremely low levels have to be detected, problems may arise with respect to sensitivity. In this case, apart from concentrating the analyte prior to ELISA, another possibility for enhancing the sensitivity of the procedure is the introduction of amplification systems, each of these introducing an enhancement of the signal per ligand. Several systems are described for this purpose hitherto, such as (i) the peroxidase-antiperoxidase system (e.g. Porstmann et al.[12]), (ii) a substrate amplification system (e.g. Stanley et al.[13], and Carr et al.[14]), and (iii) the streptavidine-biotine system (e.g. Wilchek and Bayer[15], and Guesdon et al.[16]).

Application for veterinary drug analysis. For the detection and/or determination of veterinary drug residues, various ELISAs have been described, such as methods for sulphamethazine in blood[17,18], in tissue[19] and in milk[20], for bovine somatotropin in blood and milk[21], for cephalexin in milk, tissues and eggs[22], for chloramphenicol in milk[23,24,25] and in tissues[25,26,27] and for monensin in urine and feces[28]. In some cases immunochemical methods have been used for screening purposes only. A very simple test is the test strip immunoassay (La Carte[R] test) which has been applied in the detection of chloramphenicol in urine[29], in eggs[30] and in milk[31].

The majority of these methods uses polyclonal antibodies and does not need any amplification system, as the ELISA developed was sensitive enough to detect the level required. This, however, was not the case in the ELISA for chloramphenicol (CAP) developed at our institute[26]. A monoclonal antibody, selected on specificity, was used in a competitive ELISA in which BSA-CAP was coated to the well of a microtiter plate (Fig. 2, left). In the next step a competition between CAP and immobilized CAP for a place of the (unlabeled) antibody occurs. The bound antibody was measured by incubation with a peroxide-labeled second antibody (anti-mouse) directed against the primary antibody. The amount of enzyme label bound, which is inversely proportional to the concentration of the analyte, is visualized by means of a substrate reaction. The detection level in the ELISA was such that a clean-up and concentration step of the crude muscle extract, using solid-phase extraction[32], was required. In this way, residue levels of CAP in swine muscle tissue above 5 μg kg^{-1} could be easily quantitated.

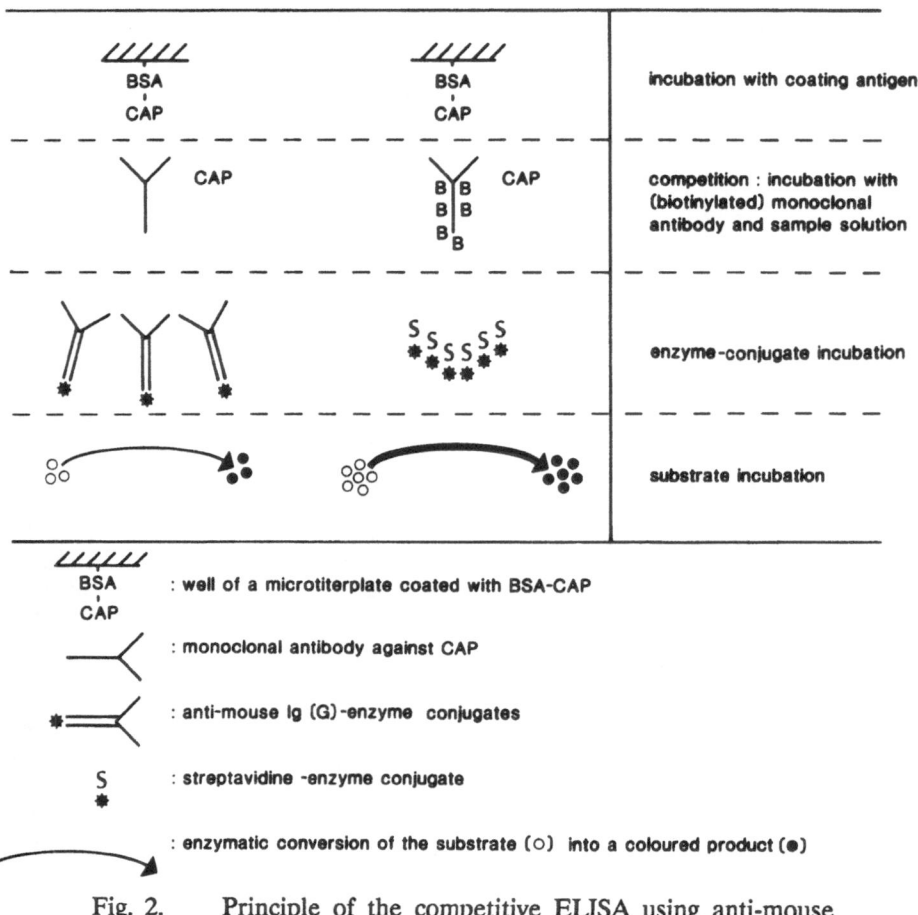

Fig. 2.　Principle of the competitive ELISA using anti-mouse Ig(G)-enzyme conjugates (left) and the modified streptavidin-biotin competitive ELISA (right) for the detection of chloramphenicol.

For direct screening of CAP in crude aqueous extracts, the detection level of the ELISA had to be lowered markedly. This was done by the introduction of a streptavidin-biotin system[27] (Fig. 2, right).

This system and the use of a coating antigen with a lower CAP incorporation resulted - after optimization - in an ELISA in which the concentration of CAP in a standard solution giving 50% inhibition decreased from 125 to 3.0 ng ml^{-1}. This resulted in the direct detection of CAP in a crude aqueous muscle tissue extract at concentrations upwards of 10 μg kg^{-1}.

It was necessary to compare the results of the ELISA of crude extracts to results obtained by analysis of a corresponding 'blank', prepared by treating a part of the aqueous sample extract with an immobilized monoclonal antibody preparation. Aqueous extracts of different swine muscle tissues show a high variation in dose-response curves, probably caused by the complexity and variability of the matrix.

The same streptavidine-biotine ELISA was also applied to the direct detection of CAP in defatted milk at a concentration of 1.0 μg kg^{-1} and higher[24].

IMMUNOAFFINITY CLEAN-UP

Another application of antibodies in residue analysis is the immunoaffinity clean-up of crude samples or sample extracts prior to a physicochemical determination. Generally the antibody, immobilized on a support, is transferred to a small column. The sample solution (sample or aqueous sample extract) is drawn through the column. This can be done by gravity flow, but also by means of a peristaltic pump. The latter ensures a constant flow through the column during loading and offers the possibility to process more columns simultaneously.

As a result of the immunochemical interactions the drug molecules are retained by the immobilized antibodies, whilst matrix components pass the column. Unbound components still remaining can be fully removed by means of a washing step. For this purpose a solvent which does not influence the immunochemical interaction, such as phosphate buffered saline (PBS), is used. After the washing procedure the antibody-bound compound is eluted. The analyte can be subsequently determined in the eluate using a suitable analytical technique such as HPLC.

The principle of this technique is not new. It has been applied for more than 25 years, in particular for isolation and purification of enzymes (e.g. Cuatrecasas et al.[33], Livingstone[34]). The use of this technique as a kind of sample preparation prior to quantitative determination, however, is a recent development[1,35,36].

Coupling. In most cases the antibodies are covalently bound to supports such as agarose, trisacryl, cellulose and polyacrylamide.

Prior to coupling the supports are activated using reagents such as cyanogen bromide (CNBr) or carbonyldiimidazole (CDI). Activated supports, e.g., CNBr-activated Sepharose (Pharmacia; Uppsala, Sweden) and CDI-activated trisacryl (Pierce; Rockford, Ill., USA) are commercially available.

The CNBr coupling procedure has some disadvantages[37]: the isourea link introduced is quite unstable, and charged isourea groups are formed which are responsible for undesirable non-specific binding due to ion exchange effects. Therefore Bethell et al.[38] and Hearn et al.[39] introduced an alternative coupling procedure using CDI. The urethane link formed in this way proved to be much more stable (leak-resistant) than the isourea link, and is uncharged.

It is important to immobilize the antibody to the support without adversely affecting the antibody's function to capture antigen. In the above-mentioned coupling procedures the covalent coupling of the protein to the carrier proceeds through the interaction of its amino groups. Generally such immunoaffinity sorbents have a lower capacity than theoretically calculated in relation to the antigen. This is mainly due to the partial inactivation of immunoglobulins during immobilization and to the shielding of the antigen-binding sites resulting from random immobilization of immunoglobulin molecules on the support[40].

Recently Matson and Little[41] have extensively studied the factors that are important for the preparation of an efficient immunosorbent. They observed that (i) monoclonal antibodies are more sensitive to coupling conditions than polyclonal antibodies, (ii) immunosorbents containing low densities of immunoglobuline G (IgG) have greater capacity for antigen on a per mole IgG basis (probably due to steric hindrance of antigen at high antibody density), and (iii) immunosorbents prepared through IgG carbohydrate linkages (oriented coupling) show large increases in antigen capacity over those prepared by stochastic (random) coupling through IgG primary amino groups. Such an oriented coupling procedure was also described by Prisyazhnoy et al.[40].

Retention and total column capacity. It is important that the conditions under which the analyte is retained on the column can be described. This, as a matter of fact, needs some definitions and quantifications.

The total column capacity is defined as the maximum amount of analyte that can be bound to the IAC column under optimal conditions. It depends on (i) the protein loading, i.e. the amount of immobilized antibodies, (ii) the volume of the immunoaffinity bed, and (iii) the orientation and affinity of the immobilized antibodies.

The affinity of antibodies can be quantified by Scatchard analysis[42] in that small molecule-protein, i.e. haptene-antibody interaction, can be described by a lineair relationship between the bound/free ratio [Ab Ag]/[Ab] and the bound fraction [Ab Ag]:

$$\frac{[Ab\ Ag]}{[Ab]} = K_a\ [q] - K_a\ [Ab\ Ag].$$

The slope of the straight line is equal to K_a, the affinity constant of the antibody. [q] is equal to the total haptene concentration.

The majority of antibodies against small molecules can be adequately described by this two-parameter model (single class of affinity).

Recently van Ginkel[36] discussed, on a theoretical base, the retention and also the elution behaviour of testosterone on an IAC column, based on the affinity constant obtained from a Scatchard plot.

The use of antibodies with a high affinity might be attractive for reasons of optimal retention. In that case, however, elution may be a problem, as either a much larger amount of eluent is required, or a more vigorous way of elution has to be applied, resulting in irreversible denaturation of antibody molecules.

The total column capacity can be determined by saturating the immunoaffinity column with the analyte. After washing the column with PBS, the analyte bound is eluted and determined. In practice, however, the amount of analyte which can be captured by the immunoaffinity column without break-through of the analyte is always smaller than the total column capacity. Apart from factors involved in the total column capacity, the capture efficiency of an antibody-mediated clean-up procedure depends on (i) the flow rate of the sample solution through the column, (ii) the concentration of the analyte in the sample solution, and (iii) the nature of the sample solution. The highest retention will be attained when small amounts of analyte molecules per time are drawn through the column (small flow in combination with a low analyte concentration) and the medium of the sample solution is close to the physiological conditions. The latter condition covers the major restriction of antibody-mediated clean-up procedures, i.e. the necessity of using aqueous extracts.

Elution. For haptene elution from the column, the conditions have to be strongly different from the conditions used for the retention. The elution procedure depends on the nature of the antigen-antibody bond, being of weak physical nature. As the types of physical bonds which are involved in immunochemical interactions vary considerably among the different antigen-antibody systems[43], each antibody-mediated clean-up procedure has its own optimal conditions for elution. Generally this elution can be based on (i) specific desorption, e.g. ligand competition and co-substrate elution, and (ii) non-specific desorption e.g. solvent changes, ionic strength alterations, buffer and/or pH changes, temperature changes, etc.[37].

Besides the desorption capacities there are several other factors which can influence the choice of the elution procedure. Apart from the compatibility of the eluent with the subsequent analysis, complete elution with mild eluents is usually desirable which makes repeated use of the column possible. Eluents able to cause irreversible denaturation of the immobilized antibodies are not preferred, as such eluents strongly restrict the repeated use of the immunoaffinity column. The latter fact is, for economical reasons, a necessity, as milligram amounts of antibodies generally are required for the preparation of one immunoaffinity column.

The factors able to improve the elution procedure are not always compatible. For example, methanol has excellent desorption properties for practical all antigen-antibody systems, so that the analyte can be eluted in a small volume. The use of methanol, however, may also result in irreversible denaturation of the antibodies which leads to reduction of column capacity. Therefore a less vigorous elution solvent has to be chosen in practice.

Application to veterinary drug analysis. Only a few applications of this sample pretreatment procedure in veterinary drug analysis have been described hitherto. These applications are on the field of anabolizing agents, ß-agonists, and chloramphenicol.

The immunoaffinity clean-up with respect to anabolizing agents was started to determine the presence of trenbolone in bovine urine by means of HPLC-TLC[44]. The polyclonal antibody used was found to bind both the active form 170-ß-trenbolone and its major metabolite 17-α-trenbolone.

A similar immunoaffinity clean-up procedure was described for the detection by HPLC and/or GC-MS of 17-ß-nortestosterone (NT) and one of its major metabolites, 17-α-nortestosterone (epiNT), in bovine urine and bile[45]. Again a single polyclonal antibody has affinity for two analytes, i.e. NT and epi-NT. In both procedures the compounds were eluted from the IAC column with a mixture of ethanol and water (40:60 v/v). Using this elution solvent, the columns could be used for many times without observing reduction in binding capacity.

Multi-immunoaffinity clean-up (MIAC), in which different polyclonal antibodies were combined in one column, was also described[36,46]. This concerns a MIAC method for the detection of (i) nortestosterone and methyltestosterone in bovine muscle tissue by means of GC/MS[46], and (ii) seven anabolics (i.e. nortestosterone, methyltestosterone, trenbolone, zeranol, diethylstilbestrol, testosterone and estradiol) in muscle tissue by means of HPLC[36]. In the first method a mixture of two different polyclonal antibodies was coupled to one support, whilst in the latter method different IAC matrices were combined in one column. Both procedures were suitable for the detection of anabolics at the low $\mu g\ kg^{-1}$ level. The mixed bed for seven anabolics was also used in an automated procedure for the detection of these seven anabolics in urine at the low $\mu g\ L^{-1}$ level[36]. Enzymatically hydrolyzed urine samples were directly injected by

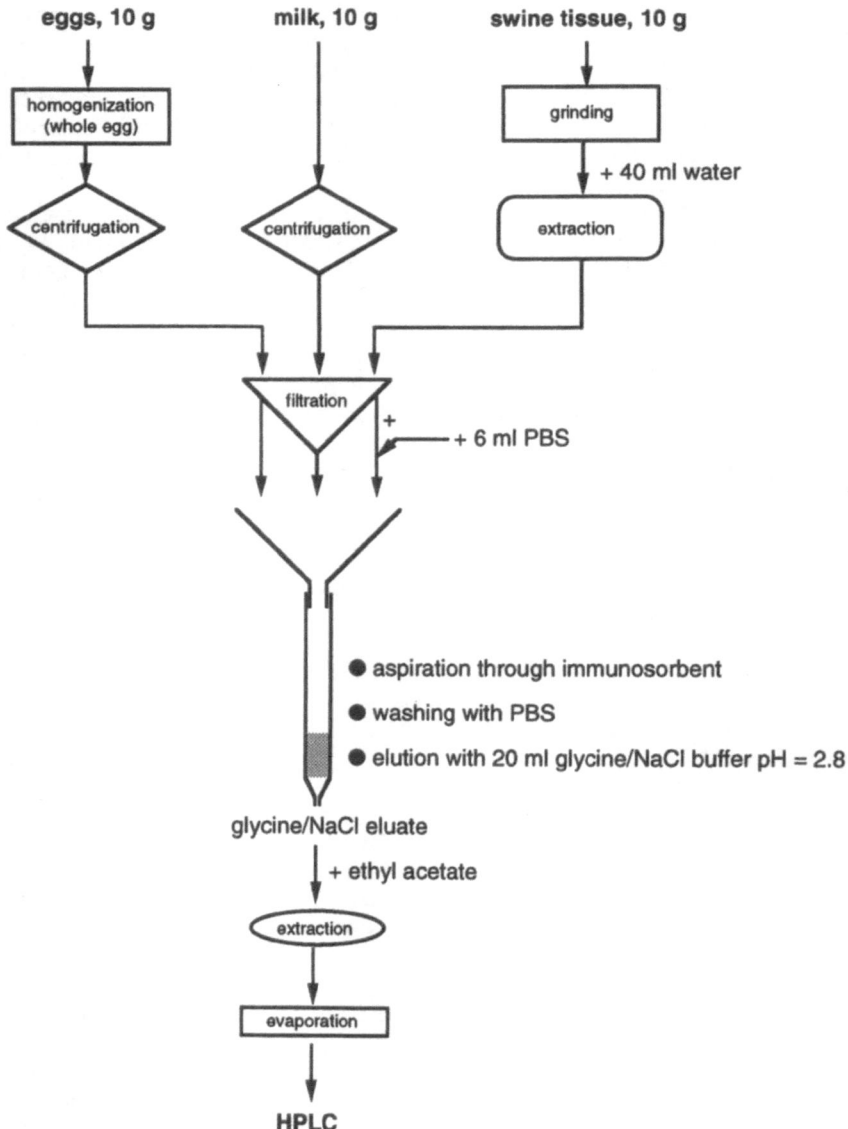

Fig. 3.   Scheme for the analysis of chloramphenicol in eggs, milk and swine tissues by means of immunoaffinity clean-up.

means of an automated injector onto the MIAC column. After elution, the analytes were trapped on a reversed-phase preconcentration column and eluted from the pre-column into the HPLC system.

An automated procedure based on the same principle was also described by Farjam et al.[47] and by Haasnoot et al.[48] for the determination of both nortestosterone and epi-nortestosterone in calf's urine and calf tissue respectively.

Recently Bagnati et al.[49] described an immunoaffinity clean-up for diethylstilbestrol and similar compounds, i.e. dienestrol and hexestrol, from urine and plasma, prior to detection by gas chromatography/negative ion - chemical ionization mass spectrometry. They applied polyclonal antibodies against DES; dienestrol and hexestrol show cross-reactions towards this antibody. The detection level was in the order of 10 ng L^{-1} (ppt). The application of a commercially available immunoaffinity column against nortestosterone (Genego; Gorizia, Italy) for the determination of 19-nortestosterone, testosterone and trenbolone by GC/MS was also described[50].

As for the ß-agonistic drugs, immunoaffinity clean-up methods based on the use of a non-specific antibody raised against clenbuterol were recently developed for the isolation of a variety of ß-agonists from aqueous samples (urine) or from cleaned meat extracts preceding detection and identification by GC/MS[36,51,52]. An on-line automated sample preparation method using immunoaffinity clean-up (IAC) in combination with a reversed-phase precolumn prior to HPLC determination of clenbuterol in urine has been also described[53].

In our laboratory, much experience was obtained with respect to the IAC of chloramphenicol from milk, egg, and crude aqueous tissue extracts[54,55,25]. The IAC columns were prepared by using monoclonal antibodies against CAP, which were originally developed for the determination of CAP residues in swine muscle tissues using an ELISA. The monoclonal antibodies were selected for specificity against CAP. Therefore the cross-reactivity towards other antimicrobial agents and structurally related compounds was negligible, thus permitting the development of a very specific clean-up and concentration procedure for CAP in aqueous extracts.

Originally, the antibodies were coupled to CNBr-activated Sepharose[54]. Later on, columns were prepared using a CDI-activated support, i.e. CDI-activated trisacryl GF-200, for reasons already mentioned[55,25]. The procedures are schematically given in Fig. 3.

Meat samples were extracted with water instead of the usual organic solvent, as these solvents are incompatible with the antibody-hapten interaction. As the optimal conditions for antibody-mediated extraction are those which approximate the physiological conditions, concentrated PBS was added to the filtered extract before immunoaffinity clean-up. Egg homogenates and defatted milk samples can be directly applied to the immunoaffinity clean-up column after centrifugation and filtration.

After washing with PBS, CAP was eluted with 20 ml of a glycine/sodium chloride solution (pH = 2.8). Elution can be performed much more efficiently with methanol but with this eluent the column cannot be reused for many times (Fig. 4). Because of this the less efficient low-pH aqueous solution had to be chosen.

For subsequent HPLC, this glycine/NaCl solution has to be extracted with ethyl acetate. After evaporation and reconstitution in eluent, chromatography on a RP-8 column was performed. Very clean chromatograms could be obtained (Fig. 5). Recoveries for milk and egg were nearly 100%. For meat a recovery of about 70 percent was obtained. This lower recovery was entirely due to losses caused by the aqueous extraction. Detection limits in the low μg kg^{-1} level could be obtained.

The results of the immunoaffinity clean-up procedure were compared with an earlier developed solid–phase extraction procedure[32], using milk and swine tissue samples obtained from CAP-treated animals. The results of both methods correspond well with each other[25] (Table 1 and 2).

Due to the large specificity of the monoclonal antibodies used, very low levels can be determined by simply increasing the test portion subjected to IAC, since no retention of matrix components occur. It was shown that 20 ng of CAP in a litre of milk can be determined with a recovery of 99%, subjecting 1 L of defatted milk to IAC. The chromatograms obtained after HPLC analysis were as clean as those obtained, after IAC, from 10 ml of milk with a much higher CAP content[55].

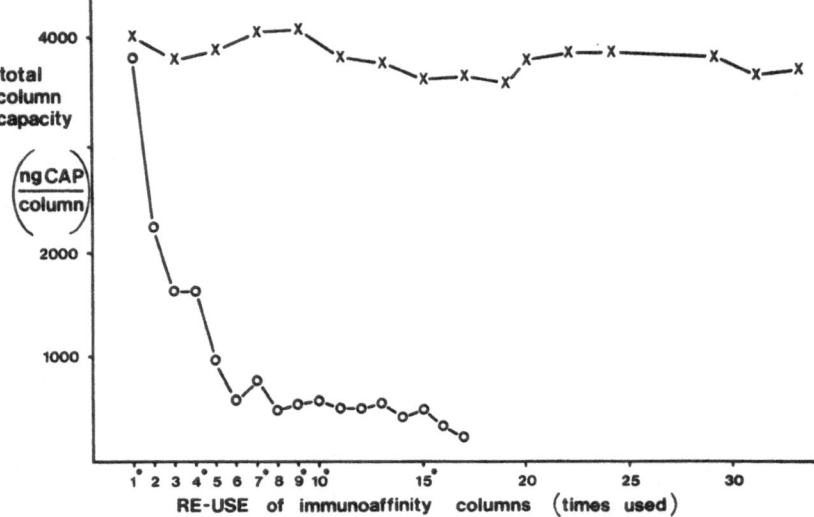

Fig. 4. The effect of the kind of eluent used on the total column capacity by re-use of immunoaffinity columns in the determination of chloramphenicol. The successive cycle, i.e. saturation, washing, elution and regeneration, was performed with two identical immunoaffinity columns (bed volume 1.5 ml). One column was eluted with methanol (o——o), whilst the other column was eluted with glycine/NaCl (x——x). The capacity determinations covered a period of one month.

Recently we applied these antibodies to an on-line high-performance liquid immunoaffinity chromatographic (HPLIAC) system for automated sample clean-up and determination of CAP in milk and swine muscle at 1 and 10 μg kg^{-1} levels, respectively[56]. HPLIAC can be considered as an integration of IAC and HPLC, in which antibodies are immobilized to an activated phase of the HPLC column.

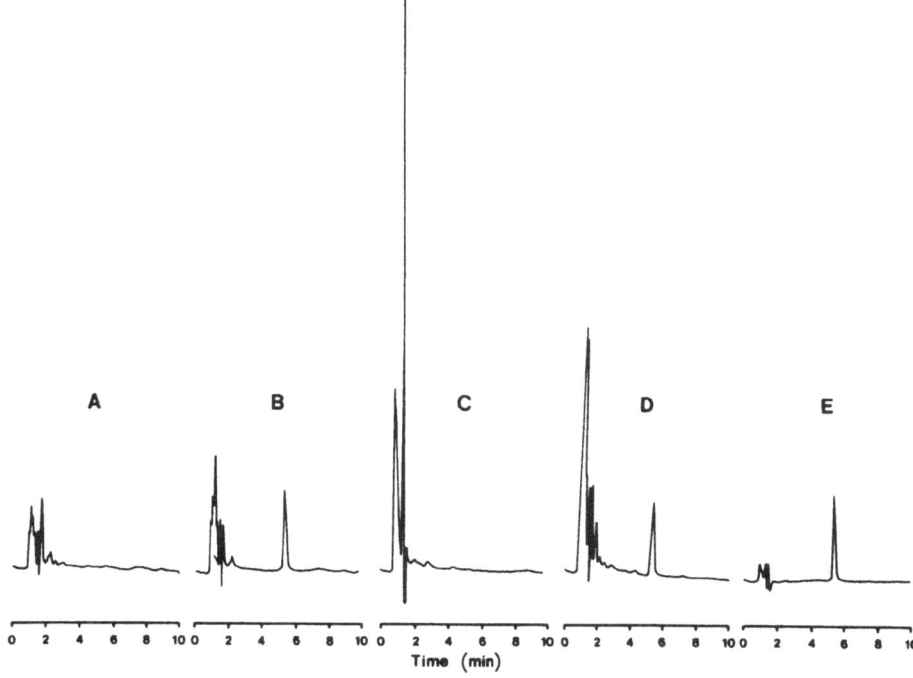

Fig. 5.   Chromatograms of milk and egg samples purified by means of immunoaffinity clean-up followed by HPLC analysis. A: blank milk sample, B: spiked (10 μg kg^{-1}) milk sample, C: blank egg sample, D: spiked (10 μg kg^{-1}) egg sample, E: standard CAP solution.

The majority of HPLIAC procedures are developed for preparative purposes (e.g. Clonis[57], Josic et al.[58], Phillips and Frantz[59]), but this technique can also be applied to analytical purposes. Such a procedure is, among others, already described for the determination of cortisol in urine[60]. In the automated procedure for CAP we coupled the monoclonal antibody to a commercially available SelectiSpher-10™ activated tresyl column (Pierce). Skimmed and deproteinized milk (5 ml) or aqueous meat extract (5 ml) were directly loaded on the HPLIAC column. After washing with PBS, CAP was desorbed by glycine/NaCl buffer (pH = 2.8). In principle, the eluate can be monitored directly by an UV detector. Due to the slow desorption, however, peak broadening occurs. Moreover, baseline stability is disturbed as a result of differences in absorption of PBS and elution solvent. Therefore we chose a dual column system in which the HPLIAC column is directly coupled to a RP-8 HPLC column. The eluate from the HPLIAC column was directly concentrated on this column. Then CAP was chromatographed with the mobile phase acetonitrile/0.01 M sodium acetate buffer (1:3 v/v).

The chromatograms of the milk samples are given in Fig. 6. No matrix interferences were observed, even at low levels. The system has been used over a three month's period in which about 150 samples were analyzed, without observing any loss in analytical performance.

Table 1. Comparison of results obtained from an IAC procedure with those from an SPE procedure using tissues and organs of a CAP-treated swine*

Name of tissue	CAP content, μg kg^{-1}	
	SPE	IAC
lean tissue		
bottom round	51	52
eye of round	52	47
top round	50	46
sirloin	49	45
loin	44	47
streaky tissue		
boston butt frontside**	3300	3600
boston butt loinside	33	32
belly	22	22
fatty tissue		
ham fat	2	2
jaw	19	17
flare	ND	ND

*   A single intramuscular injection was administered in the neck, containing 60 mg of CAP per kg body weight 64 h before slaughtering.
**   Contains injection site.
SPE = solid-phase extraction, IAC = immunoaffinity clean-up, ND = not detectable (< 1 μg kg^{-1}).

Table 2. Comparison of results obtained from an IAC procedure with those from a SPE procedure using milk of a CAP-treated cow*

Hours after injection	CAP content, μg kg^{-1}	
	SPE	IAC
0	ND	ND
15	2000	2000
23	720	680
39	61	62
47	23	27
63	6	6
73	3	2
87	2	2
95	2	1
111	ND	ND
119	ND	ND

*   A single intramuscular injection was administered in the neck, containing 30 mg of CAP per kg body weight.
SPE = solid-phase extraction, IAC = immunoaffinity clean-up, ND = not detectable (< 1 μg kg^{-1}).

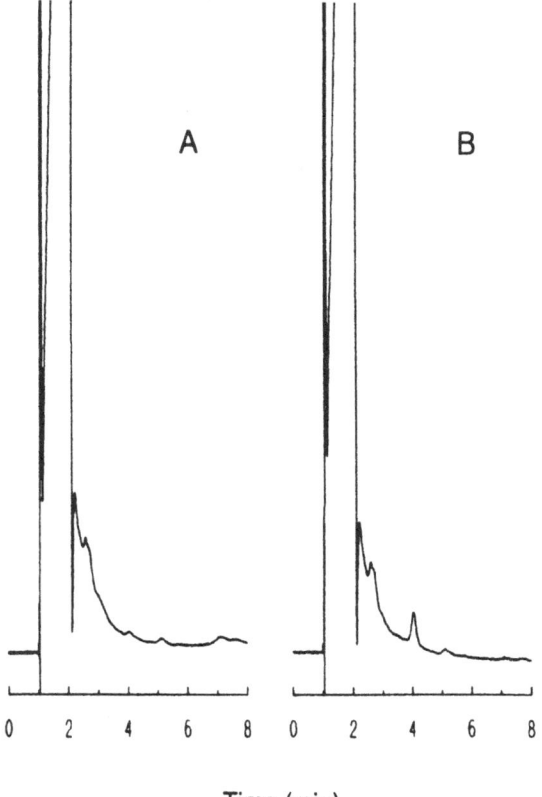

Time (min)

Fig. 6.   Chromatograms of blank (A) and spiked (B, 1 μg kg^{-1}) milk. The samples were analyzed by means of an on-line high-performance liquid immunoaffinity chromatographic system.

CONCLUSION

In recent years many developments in veterinary drug residue analysis have been made. This is also true for the immunochemical methods. Sensitive ELISAs have been developed which restrict sample treatment to a minimum, in particular when specific antibodies were applied. The sensitivity of these methods can be enhanced by the introduction of amplification systems such as the streptavidin-biotin system.

Antibodies are also successfully applied to clean-up and concentration procedures (immunoaffinity clean-up prior to a physicochemical determination). Depending on the specificity of the antibodies applied, this has led to very specific clean-up procedures as well as procedures in which a single antibody has affinity to more than one analyte. Recently, immunoaffinity clean-up procedures in which different antibodies were combined in one column were also described.

REFERENCES

1.   Haagsma, N., 1990, Sample preparation in drug residue analysis, in: "Proceedings of the EuroResidue Conference", Noordwijkerhout, The Nether-

lands, May 1990, N. Haagsma, A. Ruiter and P.B. Czedik-Eysenberg, eds., State University, Utrecht, p. 40.

2. Smith, C.J., 1990, Evolution of the immunoassay. In: Development and application of immunoassay for food analysis. J.H. Rittenberg Ed., Elsevier Applied Science, London/New York, p. 3.

3. Köhler, G., and Milstein, C., 1975, Continuous cultures of fused cells secreting antibodies of predefined specificity, Nature, 256:495.

4. Yalow, R.S., and Berson, S.A., 1959, Assay of plasma insulin in human subjects by immunological methods, Nature, 184:1648.

5. Ekins, R.P, 1960, The estimation of thyroxine in human plasma by an electrophoretic technique, Clin. Chim. Acta, 5:453.

6. Engvall, E., and Perlmann, P., 1971, Enzyme-linked immunosorbent assay (ELISA), Quantitative assay of immunoglobulin G, Immunochemistry 8:871.

7. Weemen, B.K. van, and Schuurs, A.H.W.M., 1971, Immunoassay using antigen-enzymeconjugates, FEBS Letters, 15:232.

8. Morris, B.A., and Clifford, M.N., 1985, "Immunoassays in food analysis", Elsevier Applied Science publishers, London.

9. Benraad, Th.J., Stephany, R.W., Rosmalen, F.M.A., Hofman, J.A., Loeber, J.G., and Evers, L.H., 1981, The necessity of chromatographic purification prior to radio-immunoassay of diethylstilbestrol in the urine of cattle. Vet. Q., 3:153.

10. Rattenberger, E., Matzke, P., and Neudegger, J., 1985, Entwicklung eines Radioimmunotests zur Erfassung von Rückständen des ß-Rezeptorenblockers Carazolol in Blut und Harn von Schweinen, Arch. Lebensmittelhyg., 36:85.

11. Arnold, D., Berg, D. vom, Boertz, A.K., Mallick, U., and Somogyi, A., 1984, Radioimmunologische Bestimmung von Chloramphenicol-Rückständen in Muskulatur, Milch und Eier, Arch. Lebensmittelhyg., 35:131.

12. Porstmann, T., Porstmann, B., and Seifert, R., 1983, Application of the peroxidase-antiperoxidase system as an universal reagent for the two-site binding enzyme immunoassay, Clin. Chim. Acta, 129:107.

13. Stanley, C.J., Johannsson, A., and Self, C.H., 1985, Enzyme amplification can enhance both the speed and the sensitivity of immunoassays, J. Immunol. Methods, 83:89.

14. Carr, R.I., Mansour, M., Sadi, D., James, H., and Jones, J.V., 1987, A substrate amplification system for enzyme-linked immunoassays. Demonstration of its general applicability to ELISA systems for detecting antibodies and immune complexes, J. Immunol. Methods, 98:201.

15. Wilchek, M., and Bayer, E.A., 1988, The avidin-biotin complex in bioanalytical applications, Anal. Biochem., 171:1.

16. Guesdon, J., Ternynck, T., and Avrameas, S., 1979, The use of avidin-biotin interaction in immunoenzymatic techniques, J. Histochem. Cytochem., 27:1131.

17. Fleeker, J.R., and Lovett, L.J., 1985, Enzyme immunoassay for screening sulfamethazine in swine blood, J. Assoc. Off. Anal. Chem., 68:172.

18. Singh, P., Ram, B.P., and Sharkvo, N., 1989, Enzyme immunoassay for screening of sulfamethazine in swine, J. Agric Food Chem., 37:109.

19. Dixon-Holland, D.E., and Katz, S.E., 1988, Competitive direct enzyme-linked immunosorbent assay for detection of sulfamethazine residues in swine urine and muscle tissue, J. Assoc. Off. Anal. Chem., 71:1137.

20. Dixon-Holland, D.E., and Katz, S.E., 1988, Direct competitive enzyme-linked immunosorbent assay for sulfamethazine residues in milk, J. Assoc. Off. Anal. Chem., 72:447.

21. Zwickl, C.M., Smith, H.W., and Bick, P.H., 1990, Rapid and sensitive ELISA method for the determination of bovine somatotropin in blood and milk, J. Agric. Food Chem., 38:1358.

22. Kitagawa, T., Gotoh, Y., Uchiyama, K., Kohri, Y., Kinoue, T., Fujiwara, K., and Ohtani, W., 1988, Sensitive immunoassay of cephalexin residues in milk, hen tissues and eggs, J. Assoc. Off. Anal. Chem., 71:915.

23. Märtlbauer, E. and Terplan, G., 1987, Ein enzymimmunologischer Nachweis von Chloramphenicol in Milch, Arch. Lebensmittelhyg., 38:3.

24. Water, C. van de, and Haagsma, N., 1990, A sensitive streptavidin-biotin enzyme-linked immunosorbent assay for rapid screening of residues of chloramphenicol in milk, Food Agr. Immunol., 2:11.

25. Water, C. van de, and Haagsma, N., 1991, Analysis of chloramphenicol residues in swine tissues and milk: comparative study using different screening and quantitative methods, J.Chromatogr., 566:173.

26. Water, C. van de, Haagsma, N., Kooten, P.J.S. van, and Eden, W. van, 1987, An enzyme-linked immunosorbent assay for the determination of chloramphenicol using a monoclonal antibody, Z. Lebensm. Unters. Forsch., 185:202.

27. Water, C. van de, and Haagsma, N., 1990, A sensitive streptavidin-biotin enzyme-linked immunosoebent assay for rapid screening of residues of chloramphenicol in swine muscle tissues, J. Assoc. Off. Anal. Chem., 73:534.

28. Mount, M.E., and Failla, D.L., 1987, Production of antibodies and development of enzyme immunoassays for determination of monensin in biological samples, J. Assoc. Off. Anal. Chem., 70:201.

29. Nouws, J.F.M., Beek, F., Aerts, M.M.L., Baakman, M., and Laurensen, J., 1987, Monitoring slaughtered animals for chloramphenicol residues by an immunoassay test kit (Quik-card), Arch. Lebensmittelhyg., 38:9.

30. Nouws, J.F.M., Laurensen, J., and Aerts, M.M.L., 1987, Monitoring of chloramphenicol residues in eggs by HPLC and an immunoassay (Quik-Card), Arch. Lebensmittelhyg., 38:7.

31. Nouws, J.F.M., Laurensen, J., and Aerts, M.M.L., 1988, Monitoring milk for chloramphenicol residues by an immunoassay (Quik-card), Vet. Q., 10:270.

32. Haagsma, N., Schreuder, C., and Rensen, E.R.A., 1986. Rapid sample preparation method for the determination of chloramphenicol in swine muscle by high-performance liquid chromatography, J. Chromatogr., 363:353.

33. Cuatrecasas, P., Wilchek, M., and Anfinsen, C.B., 1968, Selective enzyme purification by affinity chromatography, Proc. Natl. Acad. Sci. U.S.A., 61:636.

34. Livingstone, D.M., 1974, Immunoaffinity chromatography of proteins, in: "Methods in Enzymology", vol. XXIV, W.B. Jakoby and M. Wilchek, eds., p. 723.

35. Katz, S.E., and Brady, M.S., 1990, High-performance immunoaffinity chromatography for drug residue analysis, J. Assoc. Off. Anal. Chem., 73:557.

36. Ginkel, L.A. van, 1991, Immunoaffinity chromatography, its applicability and limitations in multi-residue analysis of anabolizing and doping agents, J. Chromatogr., 564:363.

37. Dean, P.D.G., Johnson, W.S., and Middle, F.A., eds., 1985, "Affinity chromatography, a practical approach", IRL Press Ltd., Oxford.

38. Bethell, G.S., Ayers, J.S., Hearn, M.T.W., and Hancock, W.S., 1981, Investigation of the activation of various insoluble polysaccharides with 1,1'-carbonyldiimidazole and of the properties of the activated matrices, J. Chromatogr., 219:361.

39. Hearn, M.T.W., Harris, E.L., Bethell, G.S., Hancock, W.S., and Ayers, J.A., 1981, Application of 1,1'-carbonyldiimidazole-activated matrices for the purification of proteins. III. The use of 1,1'-carbonyldiimidazole-activated agaroses in the biospecific affinity chromatographic isolation of serum antibodies, J. Chromatogr., 218:509.

40. Prisyazhnoy, V.S., Fusek, M., and Alakhov, Y.B., 1988, Synthesis of high-capacity immunoaffinity sorbents with oriented immobilized immunoglobulins or their Fab' fragments for isolation of proteins, J. Chromatogr. Biomed. Appl., 424:243.

41. Matson, R.S., and Little, M.C., 1988, Strategy for the immobilization of monoclonal antibodies on solid-phase supports, J. Chromatogr., 458:67.

42. Scatchard, G., 1949, The attractions of proteins for small molecules and ions, Ann. N.Y. Acad. Sci., 51:660.

43. Oss, C.J. van, Good, R.J., and Chaudhury, M.K., 1986, Nature of the antigen-antibody interaction. Primary and secondary bonds: optimal conditions for association and dissociation, J. Chromatogr. Biomed. Appl., 376:111.

44. Ginkel, L.A. van, Blitterswijk, H. van, Zoontjes, P.W., Bosch, D. van den, and Stephany, R.W., 1988, Assay for trenbolone and its metabolite 17α-trenbolone in bovine urine based on immunoaffinity chromatographic clean-up and off-line high-performance liquid chromatography - thin-layer chromatography, J. Chromatogr., 445:385.

45. Ginkel, L.A. van, Stephany, R.W., Rossum, H.J. van, Blitterswijk, H. van, Zoontjes, P.W., Hooyschuur, R.C.M., and Zuyderdorp, J., 1989, Effective monitoring of residues of nortestosterone and its major metabolite in bovine urine and bile, J. Chromatogr., 489:95.

46. Ginkel, L.A. van, Stephany, R.W., Rossum, H.J. van, Steinbuch, H.M., Zomer, G., Heeft, E. van de, and Jong, A.P.J.M. de, 1989, Multi-immunoaffinity chromatography: a simple and highly selective clean-up method for multi-anabolic residue analysis of meat, J. Chromatogr., 489:111.

47. Farjam, A., Jong, G.J. de, Frei, R.W., Brinkman, U.A. Th., Haasnoot, W., Hamers, A.R.M., Schilt, R. and Huf, F.A., 1988, Immunoaffinity pre-column for selective on-line sample pretreatment in high-performance liquid chromatography determination of 19-testosterone, J. Chromatogr., 452:419.

48. Haasnoot, W., Schilt, R., Hamers, A.R.M., Huf, F.A., Farjam, A., Frei, R.W., and Brinkman, U.A.Th., 1989, Determination of ß-19-nortestosterone and its metabolite α-19-nortestosterone in biological samples at the subparts per billion level by high-performance chromatography with on-line immunoaffinity sample pretreatment, J. Chromatogr., 489:157.

49. Bagnati, R., Castelli, M.G., Airoldi, L., Oriundi, M.P., Ubaldi, A., and Fanelli, R, 1990, Analysis of diethylstilbestrol, dienestrol and hexestrol in biological samples by immunoaffinity extraction and gas chromatography - negative-ion chemical ionization mass spectrometry, J. Chromatogr., 527:267.

50. Bagnati, R., and Fanelli, R., 1991, Determination of 19-nortestosterone, testosterone and trenbolone by gas chromatograpy - negative-ion mass spectrometry after formation of the pentafluorobenzylcarboxymethoximetrimethylsilyl derivatives, J. Chromatogr., 547:325.

51. Schilt, R., Haasnoot, W., Jonker, M.A., Hooyerink, H., and Paulussen, R.J.A., 1990, Determination of ß-agonistic drugs in feed, urine and tissue samples of cattle with immunoaffinity clean-up and GC-MS, in: "Proceedings of the EuroResidue Conference", Noordwijkerhout, The Netherlands, May 1990, N. Haagsma, A. Ruiter and P.B. Czedik-Eysenberg, eds., State University, Utrecht, p. 320.

52. Ginkel, L.A. van, Stephany, R.W., Rossum, H.J. van, and Farla, J., 1991, Multiresidue test for ß-agonists in a variety of matrices, in: "Proceedings of the International Symposium on Veal Calf Production", Wageningen, The Netherlands, March 1990, J.H.M. Metz and C.M. Groenestein, eds., Pudoc, Wageningen, 192.

53. Haasnoot, W., Ploum, M.E., Paulussen, R.J.A., Schilt, R., and Huf, F., 1990, Rapid determination of clenbuterol residues in urine by high-performance liquid chromatography with on-line automated sample processing using immunoaffinity chromatography, J. Chromatogr., 519:323.

54. Water, C. van de, and Haagsma, N., 1987, Determination of chloramphenicol in swine muscle tissue using a monoclonal antibody-mediated clean-up procedure, J. Chromatogr., 411:415.

55. Water, C. van de, Tebbal, D., and Haagsma, N., 1989, Monoclonal antibody-mediated clean-up procedure for the high-performance liquid chromatographic analysis of chloramphenicol in milk and eggs, J. Chromatogr., 478:205.

56. Moretti, V.M., Water, C. van de, and Haagsma, N., 1991, Automated determination of chloramphenicol in milk and swine muscle tissue using on-line immunoaffinity sample clean-up and column switching technique, in preparation.

57. Clonis, Y.D., 1987, Matrix evaluation for preparative high-performance affinity chromatography, J. Chromatogr., 407:179.

58. Josic, D., Hofmann, W., Habermann, R., Becker, A., and Reutter, W., 1987, High-performance liquid affinity chromatography of liver plasma membrane proteins, J. Chromatogr., 397:39.

59. Phillips,, T.M., and Frantz, S.C., 1988, Isolation of specific lymphocyte receptors by high-performance immunoaffinity chromatography, J. Chromatogr., 444:13.

60. Nilsson, B., 1983, Extraction and quantitation of cortisol by use of high-performance liquid chromatography, J. Chromatog. Biomed. Appl., 276:413.

RESIDUES OF ANABOLIC STEROIDS IN MUSCLE TISSUES FROM CARCASSES

WITH POSITIVE INJECTION SITES

C.H. Van Peteghem, E. Daeseleire and A. De Guesquiere

Laboratory of Food Analysis, State University of Ghent
Harelbekestraat 72, B-9000 Ghent - Belgium

INTRODUCTION

The application of anabolic steroids, although prohibited by the European Economic Community (EEC) is still a very common practice in the fattening of cattle in most of the Member-States. Each of them has established its own control apparatus in view of the compliance with the regulations.

A visual inspection of the living animal before it enters the slaughterhouse (size, behavior) may lead one to suspect the illegal administration of anabolic agents, but can never show valid evidence of illegal administration of anabolic agents. Depending on where the animal is located, different biosample types can be taken. If the animal is still at the cattle farm, urines are very easy to sample. The main advantage is that a random sampling is sufficient, as in most cases all the animals of the same species, sex and age are treated in the same way or not treated at all. In case of transgressions of the regulations, the whole stock can be put in quarantine. It can only be lifted if a re-analysis shows that all residues have disappeared and that the costs of the analyses have been paid by the owner of the stock. If on the other hand the animal has entered the slaughterhouse, the Health Inspection Services may proceed to a directed control, i.e. they may look for remnants of administered preparations, the so-called injection or implantation sites. Implantations, because they are easy to detect by a simple visual inspection of the animal or carcass, are rather seldom seen in the EEC, where every form of application of anabolic agents in view of an increased breeding efficiency is forbidden. Injection sites, if they are not too deep-lying, can easily be detected. However, the question arises whether any local inflammation is due to the injection of forbidden substances, of therapeutic agents, or simply of hypertonic saline solutions in order to mislead the control and to divert attention from the real injection site.

The injection sites can easily be removed by excision for chemical analysis. Because of the relatively high amounts of active substances present, their identification is seldom a problem for the laboratory.

As these substances are supposed to diffuse from the injection site into the blood stream of the treated animal, the question arises to what extent residues of the active ingredients will be encountered in the edible tissues of the animal. In other words, is detection of anabolic agents at an injection site an indication of consumer exposure to potentially hazardous substances?

Analysis of Antibiotic Drug Residues in Food Products of Animal Origin
Edited by V.K. Agarwal, Plenum Press, New York, 1992

In the present study an attempt is made to correlate positive injection sites to measurable residues of the parent compounds, or their metabolites, in muscle tissues. Therefore, tissue samples from 44 carcasses which had shown such a positive injection site were analyzed by means of gas chromatography-mass spectrometry (GC-MS). Concentrations in muscle tissue are in the ppb (ug/kg), or sub-ppb, range and were therefore, difficult to measure. Methods for these assays have been developed by our group (1-2).

EXPERIMENTAL

Materials and Reagents

High performance TLC-plates (silica gel 60; 10 x 10 cm), microcapillaries (Art. No. 10289), trishydroxymethylaminomethane (Tris), sulfuric acid and benzene were from Merck (Darmstadt, Germany). Diethylether, ethylacetate, chloroform, acetone, cyclohexane and ethanol were purchased from Janssen Chimica (Geel, Belgium). Methanol and water of HPLC grade were obtained from Alltech (Deerfield, IL, U.S.A.). Heptafluorobutyric acid anhydride (HFBAA) from Macherey-Nagel (Duren, Germany), N-methyltrimethylsilyltrifluoroacetamide (MSTFA) and trimethylsilyliodide (TMSI) from Aldrich (Milwaukee, WI, U.S.A.), and trimethylchlorosilane (TMCS) from Fluka (Buchs, Switzerland) and pyridine form Pierce (Rockford, IL, U.S.A.) were used as supplied. Octadecyl silane (ODS) disposable extraction columns were from J.T. Baker (Phillipsburg, N.J., U.S.A.).

The derivatization vials were silanized with a solution of dimethyldichlorosilane (Merck) in toluene (Merck) before use.

Analysis of Injection Sites

The injection site was cut in little pieces which were put in a polyethylene bag. After addition of 5 ml of methanol, the content of the bag was homogenized for 5 minutes in a blender (Stomacher Type 80, Colworth, London, U.K.). The methanol was taken off and evaporated to dryness.

The evaporated extract was taken up in 100 ul of methanol and a 0.75 ul aliquot spotted on a HPTLC-plate. Two chromatographic runs were performed in two opposite directions of one HPTLC plate. On a starting line 1 cm from one edge, one 0.75 ul aliquot of the extract was spotted with a microcapillary. On the opposite side of the plate, again 1 cm from the edge, the spotting was repeated. The reference standards were spotted in groups in six different lanes also on both starting lines.

The plate was eluted in one direction till the middle of the plate with chloroform-acetone (90:10 V/V). After drying, the opposite half of the plate was eluted with cyclohexane-ethyl acetate-ethanol (77.5:20:2.5 V/V/V).

After drying, the chromatogram was sprayed with 10% sulfuric acid in methanol and heated for 10 minutes at 90^{0}C. The spots were examined under UV light at 366 nm and in visible light.

Analysis of Muscle Tissue

Each muscle tissue sample (1 or 2 g) was digested overnight at 60^{0}C in 4 ml of a 0.1 M Tris solution (pH 9) containing 1 mg of the proteolytic enzyme Subtilisin A. The digest was then extracted with two 5 ml volumes of diethylether, and the combined fractions were evaporated to dryness. The residue was taken up in 0.1 ml of methanol to which 4 ml of water were added.

This solution was applied onto a C18 solid phase extraction (SPE) column, previously conditioned with 2x3 ml of methanol and 2x3 ml of water. After washing the column with 2x3 ml of water, the analytes were eluted with 2 ml of methanol. After evaporation to dryness, the residue was taken up in 100 ul of methanol and a 50 ul portion was subjected to an HPLC fractionation step.

HPLC Fractionation Step

Fractionation was carried out with an HPLC system consisting of a system controller and pump (model 600E), an automatic injector (WISP 710 B), an UV detector (model 484), 2 automatic switching valves, all from Waters (Milford, MA, U.S.A.) and a fraction collector model 2212 Helirac (Pharmacia-LKB, Uppsala, Sweden). The analytical column was a Lichrospher RP 18 (125 mm x 4 mm I.D., 5um) (Merck) and was protected by a guard column (pellicular reversed phase 30-50 um, 75 mm x 2.1 mm I.D.) (Chrompack Cat. No. 28306, Middelburg, The Netherlands). The mobile phase was methanol-water (65:35, V/V) and was pumped at a flow rate of 1 ml/min. The UV detector was set at 254 nm.

The methanol- water fractions were evaporated to dryness under a stream of nitrogen.

The collection windows were determined by injecting 25 ng of each standard with the UV detector set at the wavelength of the maximum absorption of the compound.

Derivatization

Heptafluorobutyrates (HFB). The purified extract obtained from the HPLC fractionation step was dissolved in 0.2 ml of benzene and 0.05 ml of HFBAA were added. The mixture was heated for one hour at 60^0C and then concentrated to dryness at 40^0C under a flow of nitrogen. The final residue was redissolved in 50 ul of hexane, 3 ul of which were injected into the GC-MS instrument.

Trimethylsilyl (TMS) derivatives. The purified extract obtained from the HPLC fractionation step was treated with 100 ul of pyridine-MSTFA-TMCS (10:3:1, V/V/V) and heated for one hour at 60^0C. After evaporation of excess reagent, the final residue was dissolved in 50 ul of hexane and a 3 ul portion was injected into the GC-MS instrument.

In some cases, and particularly when the molecule contains a 3 keto-function, an enolization is preferable. In those cases, 50 ul of MSTFA-TMSI (1000:2, V/V) was added, the mixture heated for 30 minutes at 60^0C and injected without further treatment.

Gas Chromatography-Mass Spectrometry (GC-MS)

Analyses were carried out on an HP 5970 mass selective detector (Hewlett Packard, Palo Alto, CA, U.S.A.) linked to an HP 5890 gas chromatograph equipped with an HP Ultra 2 (5% phenylmethyl silicon) fused silica capillary column (25 m x 0.2 mm I.D., film thickness 0.33 um) and an all glass moving-needle injection system. The carrier gas was high-purity helium (L'air liquide, Liege, Belgium) at a flow rate of 0.47 ml/min. The injector and interface temperature were maintained at 290^0C. The oven temperature was programmed from 200 to 280^0C at 5^0C/min., the final temperature being held for 10 minutes.

RESULTS

The anabolic agents which are more or less likely to be encountered in injection solutions are shown in Table 1. In view of the thin-layer chromatographic analysis they are classified in six groups so that all the compounds within one group are completely resolved. Color differentiation may help when interfering substances yield spots with the same RF value.

A particular compound is considered not to be present if it is not unambiguously identified in both chromatographic runs.

A compound is considered to be identified by gas chromatography-mass spectrometry if it appears in the gas chromatogram at the right retention time, i.e. the retention time expressed in methylene unit values of a reference standard run under the same conditions, and if all the selected ions, as indicated in Table 2, simultaneously appear at the correct retention time.

Although if was prescribed by the EEC (3) that the relative abundances of all the selected ions monitored for the analyte should match those of the reference standard, practical experience has evidenced that this is merely a theoretical criterion. Indeed, blank samples to which the analyte was added by spiking prior to extraction, in some cases should have been declared negative if those rules were to be applied too stringently.

Between January 1, 1990 and August 15, 1991, 608 injection sites were analyzed. Out of these, 369, or 60.7%, contained at least one forbidden anabolic agent. Details as to the frequency distribution of the individual agents are shown in Table 3.

With the collaboration of the Public Health Inspection Services, a muscle sample was taken in 44 cases of a positive injection site. Attention was paid that the sampling was not done in the vicinity of the excised injection site. The correlation between the findings in the injection site and in the muscle tissue is shown in Table 4.

DISCUSSION

A tentative conclusion which can be drawn from this experiment is that the presence of an injection site in a carcass does not necessarily mean that there are substantial residue concentrations in the edible tissues. The limit of detection by GC-MS, for most of the compounds studied, is in the order of 0.2 ppb. Factors contributing to the absence of residues may be: 1) dose, 2) time elapsed since administration (encapsulation of the oily solution), 3) sample type, 4) distance between injection and sample site, and 5) pharmaceutical formulation.

Some of the anabolic agents, which are known to be fat-soluble, because they occur in the form of acetate esters that are not hydrolyzed in the body, seldom or never yield residues. This is the case for chlorotestosterone acetate, medroxyprogesterone acetate and chlormadinone acetate. These lipophilic substances dissolve in the lipid fraction of the muscle tissue extract, which is being discarded during the extract purification.

Finally, a hypothesis, which can be set up based on some recent literature data (4-8), is that phenolic groups of free steroids may be esterified with fatty acids in the animal, as a result of which they become lipophilic and behave like the non-hydrolyzed acetate esters.

All these hypotheses are the subject of an ongoing study in the author's laboratory.

Table 1. List of anabolic steroids included in the HPTLC detection system.

Group No. 1	Dienestrol
	Estradiol benzoate
	Mestranol
	Trenbolone acetate
	4-Chlorotestosterone acetate
	19-Nortestosterone decanoate
	Chlorotestosterone
Group No. 2	Estradiol 17β
	Diethylstilbestrol
	Medroxyprogesterone acetate
	Estradiol 17β-cypionate
	Norethisterone acetate
	Testosterone cypionate
Group No. 3	19-Nortestosterone
	17-α-Methyltestosterone
	Megestrol Acetate
	Testosterone isocaproate
	19-Nortestosterone laurate
	Progesterone
Group No. 4	Boldenone
	Diethylstilbestrol dipropionate
	Chlormadione acetate
	Estradiol 17β- fenylpropionate
	Testosterone enantate
	Testosterone propionate
Group No. 5	Stanozolol
	Testosterone
	Ethinylestradiol
	19-Norethisterone
	Estradiol valerate
	Testosterone fenylpropionate
Group No. 6	Methandienone
	Zeranol
	Hexestrol
	Vinyltestosterone
	Melengestrol acetate

Table 2. Data of the HFB and TMS derivatives.

	HFB Derivatives			TMS Derivatives		
	M.U.V	M+	Selected ions	M.U.V	M+	Selected ions
17α-19-Nortestosterone	23.65	666	666,453,306	26.36	418	418,194,182
17β-19-Nortestosterone	24.40	666	666,453,306	26.84	418	418,194,182
17α-Estradiol	23.97	664	664,451,409	26.68	416	416,285,129
17β-Estradiol	24.66	664	664,451,409	27.28	416	416,285,129
17α-Testosterone	23.85	680	680,467,320	27.73	432	432
17β-Testosterone	24.62	680	680,467,320	27.28	432	432
Ethinylestradiol	25.87	492	492,409,356	28.42	440	440,425,285
Trenbolone	24.71	662	662,448	30.03	413	413,338
Methyltestosterone	22.72	694	480,465,369	28.26	446	446,356,301
Medroxyprogesterone acetate	29.74	582	582,479,383	–	–	–
Clormadinone acetate	30.55	497	497,461	–	–	–
Megestrol acetate	29.07	580	580,477	–	–	–
Diethylstilbestrol	20.32	660	660,447,341	23.91	412	412,397,383
Hexestrol	20.43	662	449,331,303	23.76	403	399,207
Dienoestrol	22.38	656	656,629,445	23.88	410	410,395
Zeranol	22.82	910	696,555,358	28.69	538	538,523,433
Methylboldenone	22.45 23.86	478	478,463,367	28.82	444	444,354,339
Boldenone	23.66 25.32	678	678,464,343	27.89	430	430,325,299
Progesterone	26.26 26.88	510	510,425	29.5	458	458,443,353
4-Chlorotestosterone	28.81	518	518,482,441	29.74	466	466,431,451
Chlorotestosterone acetate	–	–	–	30.60	436	436,421,401

M.U.V. : Methylene Unit Value
M+ : Molecular Ion

Table 3. Occurrence of anabolic agents in 369 positive injection sites.

	No. of times encountered in 369 positive injection sites	% frequency in posi- tive injec- tion sites
Testosterone esters	260	70.5 %
Estradiol 17β-esters	256	69.4 %
Estradiol 17β	244	66.1 %
Testosterone	236	64.0 %
Chlorotestosterone acetate	154	41.7 %
19-Nortestosterone esters	132	35.8 %
19-Nortestosterone	109	29.5 %
Methyltestosterone	92	24.9 %
Medroxyprogesterone acetate	76	21.5 %
Stanozolol	75	20.3 %
Progesterone	41	11.1 %
Boldenone	23	6.2 %
Ethinylestradiol	21	6.0 %
Megestrol acetate	18	5.1 %
Chlorotestosterone	6	1.7 %
Methylboldenone	5	1.4 %
Chlormadinone acetate	5	1.4 %
Trenbolone acetate	1	0.3 %

Table 4. Correlation of findings in injection sites and muscle tissue.

Anabolic agent	No. of times encountered in 44 positive injection sites	No. of times present in corresponding muscle tissue
Estradiol 17β and esters	20	10
Testosterone and esters	18	8
19-Nortestosterone and esters	18	4
Methyltestosterone	17	4
Chlorotestosterone acetate	11	0
Medrozyprogesterone acetate	11	1
Stanozolol	9	0
Estradiol 17β esters	8	2
Megestrol acetate	7	0
19-Nortestosterone esters	6	4
Progesterone	5	0
Testosterone	4	1
Chlormadinone acetate	4	0
Estrdiol 17β	3	0
Ethinylestradiol	3	1
Testosterone esters	2	1
Trenbolone acetate	1	0

ACKNOWLEDGMENT

This work was supported by IWONL (grant no. 5265A), The Institute for Veterinary Control of the Ministry of Public Health and The National Fund for Scientific Medical Research (grant no. 3.0017.89).

REFERENCES

1. E. Daeseleire, A. De Guesquiere and C. Van Peteghem, Derivatization and GC/MS Detection of Anabolic Steroid Residues Isolated from Edible Muscle Tissues, J. Chromatogr. 562:673 (1991).

2. E. Daeseleire, A. De Guesquiere and C. Van Peteghem, Multi-residue Analysis of Anabolic Agents in Muscle Tissues and Urines of Cattle, Anal. Chem. (submitted).

3. Commission Decision of 14 July 1987 laying down the methods to be used for detecting residues of substances having a hormonal action and of substances having a thyrostatic action, Off. J. Eur. Commun. L223:18 (1987).

4. D.L. Jones, and V.H.T. James, The Identification, Quantification and Possible Origin of Non-polar Conjugates in Human Plasma, J. Steroid Biochem. 22:243 (1985).

5. L. Janocko and R.B. Hochberg, Radiochemical Evidence for Estradiol-17-fatty Acid Esters in Human Blood, J. Steroid Biochem. 24:1049 (1986).

6. J.B. Adams, P. Martyn, D.L. Smith and S. Nott, Formation and Turnover of Long-chain Fatty Acid Esters of 5-androstene-3B, 17B-diol in Estrogen Receptor Positive and Negative Human Mammary Cancer Cell Lines in Culture, Steroids 51:251 (1988).

7. R. Roy, and A. Belanger, Formation of Lipoidal Steroids in Follicular Fluid, J. Steroid Biochem. 33:287 (1989).

8. A. Paris and D. Rao, Biosynthesis of Estradiol-17B Fatty Acyl Esters by Microsomes Derived from Bovine Liver and Adrenals, J. Steroid Biochem. 33:465 (1989).

APPROACHES TO THE DETECTION AND CONFIRMATION OF DRUG RESIDUES IN MILK

Mary C. Carson, David N. Heller, Philip J. Kijak, and
Michael H. Thomas

Division of Veterinary Medical Research, Center for
Veterinary Medicine, U.S. Food and Drug Administration,
Beltsville, MD 20705

INTRODUCTION

The possible presence of trace residues of veterinary drugs in milk is
a subject of major concern for regulatory agencies, the dairy industry, and
consumers. All New Animal Drug Applications (NADAs), in addition to
showing efficacy and safety of the drug to the animal, must also show that
treatment of the animal will result in no hazardous residues entering the
human food chain. NADAs therefore include methodology to detect tissue
residues. Part of the review process at FDA's Center for Veterinary
Medicine (CVM) is to evaluate the drug sponsor's residue method.

For several reasons, CVM also has its own method development program.
Many veterinary drugs were approved prior to current regulatory requirements. For these older drugs, sensitive analytical methodology may not
exist for residues in milk. Another problem is extra-label use of approved
drugs. New drugs are approved for use in particular species for certain
conditions; the label may even specify the age and sex of the animal. Very
few drugs are actually approved for use in lactating dairy cattle. Careless use of drugs approved for other animals (e.g., calves), for which the
milk withholding time for lactating cows is not known, may result in residues. Finally, the use of some drugs, such as chloramphenicol, has been
completely banned in food producing animals. Sensitive analytical methods
are needed to detect illegal use of these drugs.

For regulatory purposes, a residue method must have a high degree of
both precision and specificity. In actual practice, this usually means two
different types of analytical procedures, which together comprise the
method. These analytical assays are referred to as determinative and
confirmatory procedures.

A determinative procedure accurately quantitates residue levels in
tissues. The procedure should be able to accurately determine residues at
half the tolerance or concern level. For residue levels less than 0.1 ppm,
the recovery from fortified tissue should be > 60% and the coefficient of
variation < 20%. Interferences should be less than 10% at the residue
concentration of interest. Finally, the procedure should be practicable,
both in terms of time and resources needed. An experienced chemist should
be able to process at least 6 to 8 (FDA will probably soon change this

Analysis of Antibiotic Drug Residues in Food Products of Animal Origin
Edited by V.K. Agarwal, Plenum Press, New York, 1992

107

number to 12) samples a day using equipment commonly available in an analytical lab.

Most chemical determinative procedures are chromatographic. A peak with a given retention time is not considered specific enough to confirm the identity of a residue. A confirmatory procedure may be a second chromatographic assay using entirely different separation conditions. However, preferred confirmatory procedures generally employ mass spectral analysis of the residue, with identification of four structurally significant fragment ions, or three ions if one is the molecular ion. FDA considers this sufficient to confirm presence of the residue. Confirmatory procedures should be able to detect residues at the concern level.

Goals of CVM method development are several-fold:

1) Lower the detection limits of existing milk methods to meet current regulatory concern levels.
2) Adapt existing tissue residue methods for use in milk.
3) Decrease reliance on organic solvents, especially halogenated solvents, in the extraction procedure. Proper disposal of hazardous waste is expensive and adds significantly to the cost of an assay. It may also result in unnecessary analyst exposure.
4) Develop multiresidue methods. Frequently, several members of a class of drugs are commercially available, and any or all of these drugs may be used in veterinary practice. Methods that detect all or most members of a drug family may be more difficult to develop, but will save time and resources in actual field use.

We briefly describe here three procedures either recently developed, or currently being developed at CVM, which illustrate these points.

LC DETERMINATIVE PROCEDURE FOR DETECTION OF TETRACYCLINES IN MILK

Introduction

Tetracyclines (TCs) are a class of antibiotics containing a partially conjugated four ring structure. Oxytetracycline (OTC), tetracycline (TC), and chlortetracycline (CTC) are available in veterinary formulations. Table 1 lists these drugs, their tolerance levels in milk as published in the *Code of Federal Regulations* (CFR)[1], FDA's concern level (level at which regulatory action will be taken), and approved uses and withdrawal times in lactating dairy cattle.

In addition to these three major TCs, several other TCs are available commercially which we decided to include in this study. These are minocycline (mino), demeclocycline (DMCTC), methacycline (metha), and doxycycline (doxy). These drugs are available primarily for human use, though doxy is also approved for use in dogs.

For the purposes of this study, a target marker residue concentration of 30 ppb was selected for all seven TCs. To satisfy the above criteria for a determinative procedure, this meant that the procedure developed would need to be sensitive to at least 15 ppb. This level is that required for OTC and CTC, but is actually more sensitive than required for TC. Initial studies showed that the gradient LC conditions used by Thomas[2] adequately resolved all seven TCs. However, interferences in the milk limited the sensitivity of the Thomas procedure to about 100 ppb.

Use of matrix solid-phase dispersion (MSPD)[3] was also not sufficient to achieve the sensitivity we desired. At this point, we tried metal

Table 1. Tetracyclines in Veterinary Use

Drug	CFR Tolerance Level in Milk	FDA Concern Level	Approved Uses in Lactating Cows	Discard Time
OTC	none	30 ppb	Intra-mammary infusion Feed	4 days None
TC	none	80 ppb	none	--
CTC	zero	30 ppb	Feed	None

chelate affinity chromatography (MCAC). The principle of MCAC for use in detecting OTC, TC, and CTC in tissue was originally developed by Farrington, et al.[4,5] Our procedure is a substantial modification of his, enabling processing of more samples per day and minimizing the amount of organic solvent consumed (only 2 mL methanol are used in the extraction).

Experimental

Milk (5 mL) was centrifuged to effect separation of the cream. The "skim" milk was acidified with pH 4 succinate buffer to precipitate most of the proteins, and the clarified milk passed through a mini-column containing approximately 1 mL of Chelating Sepharose Fast Flow (Pharmacia) which had been previously "charged" with Cu^{2+}. The mini-columns were washed with buffer, water, and methanol, and then the absorbed TCs eluted with EDTA-containing buffer. The eluate was ultrafiltered to remove remaining

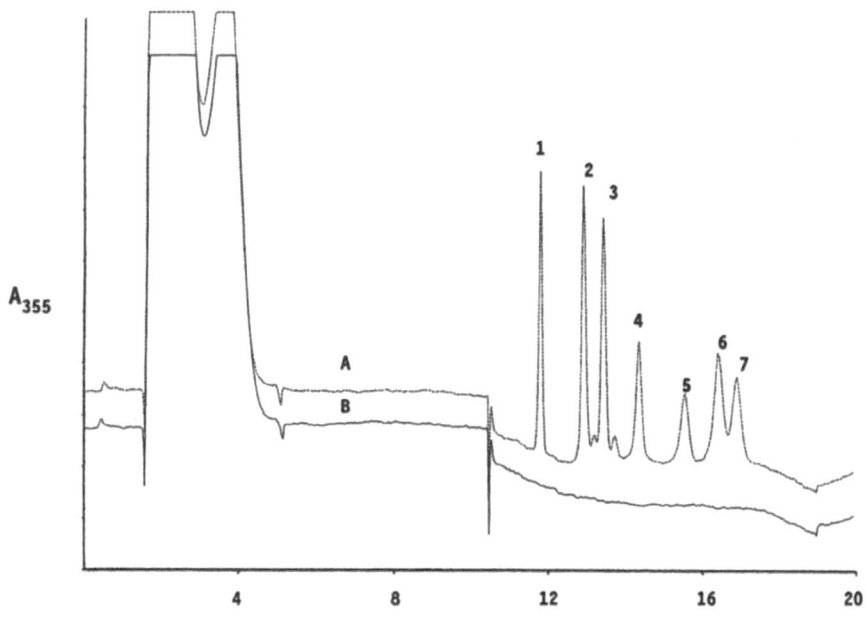

Fig. 1. Chromatographs of MCAC extracts of (A) control milk fortified to 30 ppb each mino (1), OTC (2), TC (3), DMCTC (4), CTC (5), metha (6), and doxy (7) and (B) control milk.

Table 2. Recoveries of TCs from Fortified Milk Samples

Fortified Sample Concentration	Recovery (percent) with CV (percent)						
	Mino	OTC	TC	DMCTC	CTC	Metha	Doxy
15 ng/mL	66.6	91.0	79.4	81.0	80.3	74.6	79.3
n = 5	8.9	9.5	10.7	12.0	21.3	16.7	13.0
30 ng/mL	65.1	85.6	73.6	75.2	68.3	67.1	72.4
n = 6	12.2	8.0	4.5	2.8	9.5	6.7	6.0
60 ng/mL	63.4	82.0	69.8	68.6	65.5	65.3	71.6
n = 6	13.5	10.1	5.5	4.7	6.5	6.1	6.4

proteins, and the filtrate injected directly onto the LC. LC conditions were as described by Thomas[2], except that the column used was a 4.6 x 150 mm PLRP-S column (Polymer Labs) rather than the Waters Nova-Pak column.

For validation of the procedure, control milk was collected from drug-free cows and fortified with TCs at 15, 30, and 60 ppb for recovery studies. Cows were also treated with oral doses of each of the seven TCs to generate milk containing incurred residues. This milk was used for reproducibility and exhaustive extraction studies.

Results

The MCAC process produces an extract essentially free of endogenous milk peaks, allowing a very low (< 5 ppb) limit of detection for most TCs. Fig. 1 shows chromatograms of control milk and milk fortified with 30 ppb of each of the seven TCs. The number of samples which can be processed in one day is limited only by the thermal stability of the TCs themselves. Typically, 14 samples are processed during the day and then analyzed overnight using an LC equipped with an autosampler. TC, and especially CTC, start to degrade significantly after 4 hours at room temperature, limiting the number of samples which can be loaded in an unrefrigerated autosampler. Processed samples may be stored refrigerated 24 hours or more with no significant loss of TCs.

Table 2 shows the recovery data from the fortified milk samples. In all instances, the average recovery exceeded 60%. Reproducibility studies indicated that the coefficient of variation in incurred samples was < 10% for intra-day experiments and < 20% for inter-day experiments. Exhaustive extraction of OTC, TC, and CTC incurred milk showed that 88%, 92%, and 90%, respectively, of the total extracted residues were extracted in the procedure as developed. Allowing for slight losses during the washing of the MCAC columns, these numbers are consistent with the recovery data from the fortified milk samples.

The performance of this procedure has been verified by having a second analyst in our laboratory use it to assay for tetracyclines in milk. We are currently undergoing preparations for an inter-laboratory method trial.

GC/MS CONFIRMATORY PROCEDURE FOR CHLORAMPHENICOL IN MILK

Introduction

Chloramphenicol (CAP) is a broad spectrum antibiotic known to cause aplastic anemia in some humans. Due to health concerns, the World Health

Organization recommended its use not be allowed in food producing animals.[6] The FDA has banned the use of CAP in all food producing animals. A wide variety of methods have been developed for the detection of CAP in milk. Allen reviewed methods for the detection of CAP in milk and meat.[7] More recently, the use of diatomaceous earth-filled solid phase extraction (SPE) columns has been explored for the isolation of CAP in milk.[8] However, these methods lack the sensitivity and specificity required of a confirmatory procedure. A procedure capable of confirming the presence of CAP residues in meat, based on gas chromatography/negative chemical ionization-mass spectrometry (GC/NCI-MS), has been developed by the Food Safety and Inspection Service of the United States Department of Agriculture.[9]

The procedure described here is able to confirm the presence of 0.5 ng/mL of CAP in milk by gas chromatography/mass spectrometry. NCI with selected ion monitoring is used to determine the presence of CAP in the sample. Sample preparation is based on SPE to minimize solvent use, with meta-nitrochloramphenicol (m-CAP) added as a surrogate to demonstrate procedure performance.

Experimental

The surrogate was added to 10 mL of milk. The samples were mixed with ethyl acetate, loaded onto a diatomaceous earth-filled SPE column, and the CAP eluted from the column using more ethyl acetate.[8] A 4% NaCl solution was added to the sample extract, and the ethyl acetate removed by evaporation. Following the removal of remaining fat with hexane, the 4% NaCl solution was loaded onto a C-18 SPE column, the column washed with water, and methanol used to elute the CAP. After the methanol was evaporated, the trimethylsilyl derivative of CAP was formed. The excess derivatizing reagent was evaporated, and cyclohexane/hexane used to dissolve the sample for analysis by GC/NCI-MS.

Confirmation (CS) was based on having a minimum of five of the six monitored ions for chloramphenicol present, and the isotopic and/or fragment ion relative abundances within 10 percentage units of the values obtained using standards. For example, if a fragment ion had a relative abundance of 20% in a standard, the relative abundance of that fragment ion in a sample would have to be between 10% and 30% for confirmation. An ion was present if the peak for the ion was greater than the limit of detection as defined by Keith, et al.[10] The values for standards were obtained by averaging the values from the analyses of standards prepared with each set of samples. Samples that did not meet the requirements were considered failed to confirm (FTC). The criteria used for confirmation are within established guidelines for regulatory methods.[11,12] Any sample that had a meta-nitrochloramphenicol peak < 10% of that found in standards was considered unsuccessfully extracted and was not assigned FTC status.

Results

Data for the validation of the method were generated by analyzing a series of 42 samples consisting of 7 control, 21 fortified, and 14 milk samples from a cow dosed with CAP. Selected ion chromatograms for control milk and milk containing incurred residues are shown in Fig. 2.

Table 3 summarizes the results of the analyses of the validation samples. All 7 control milk samples were assigned FTC. The presence of CAP was confirmed in 20 of the 21 fortified milk samples and all 14 dosed milk samples. One 0.5 ng/mL CAP fortified sample could not be assigned a status. Only three ions were present and the amount of m-CAP measured in this sample was less than 10% of that found in the standards, indicating poor recovery.

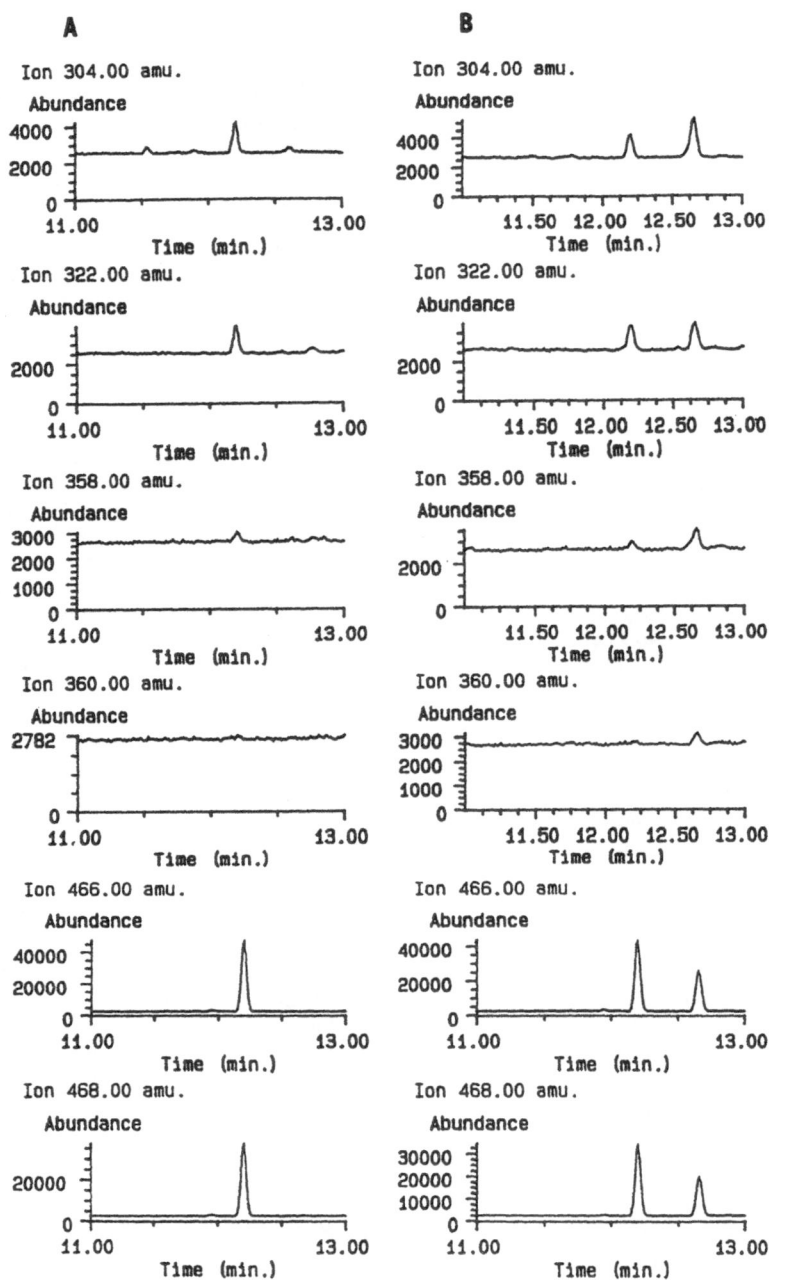

Fig. 2 Selected ion chromatograms of A) control milk and B) incurred milk extracts. Peaks at 12.2′ are from surrogate *m*-CAP. Peaks at 12.7′ derive from CAP.

Table 3. Summary of Status of CAP Confirmatory Procedure Validation Samples

Sample number	1	2	3	4	5	6	7
Control milk	FTC	FTC	FTC	FTC	FTC	FTC	FTC
0.5 ng/mL Fortified milk	CS	CS	--	CS	CS	CS	CS
1.0 ng/mL Fortified milk	CS	CS	CS	CS	CS	CS	CS
2.0 ng/mL Fortified milk	CS	CS	CS	CS	CS	CS	CS
Incurred milk 736 hour post dose	CS	CS	CS	CS	CS	CS	CS
Incurred milk 928 hour post dose	CS	CS	CS	CS	CS	CS	CS

This procedure was evaluated during development to identify critical steps and parameters, and insure that the procedure would be suitable for the analysis of regulatory samples. This evaluation included performing a ruggedness test as described by Youden.[13] Several critical areas were identified. These areas included choice of C-18 SPE columns, glassware cleaning, water purity, and the completeness of the sample drying before derivatization. In addition, the factors chosen to confirm CAP in a sample were evaluated to determine if they were sufficiently selective without being overly restrictive. Finally, the performance of this procedure was checked by having a second analyst in our laboratory use the procedure to confirm the presence of CAP in milk samples.

DEVELOPMENT OF A PROCEDURE FOR CONFIRMATION OF IVERMECTIN IN MILK BY LC/MS

Introduction

Ivermectin is the commercial name for an anti-parasitic medication consisting of a mixture of two structural homologs, dihydro-avermectin B_{1a} and B_{1b} (H_2B_{1a}, H_2B_{1b}), in at least an 80:20 ratio. It is presently not approved for dairy cattle. The H_2B_{1a} homolog is the marker residue for ivermectin and serves as the basis for determinative and confirmation procedures. Its structure consists of a 16-membered lactone ring and two sugar substituents, with a molecular weight of 874 (Fig. 3). At present, there is no approved confirmatory procedure for ivermectin in milk. A confirmatory procedure has been approved for animal tissue, based on partial hydrolysis of ivermectin to its aglycone and monosaccharide, followed by conversion to fluorescent derivatives and LC analysis with fluorescence detection of the three peaks. However, this method does not use mass spectrometry, and in practice it is similar to the approved determinative method, which has been described as lengthy and tedious.[14]

A recent depletion study of ivermectin in dairy cows suggested that considerable excretion occurs via the mammary gland,[15] underscoring the possibility that residues might occur in milk. Ivermectin is administered at low levels due to its potency, so a confirmatory procedure must show good sensitivity. An extrapolation of the concern levels for muscle and liver tissue translates into a detection goal of 2 ppb in milk. This level of sensitivity, equivalent to 20 ng in 10 mL of milk, had not previously been demonstrated for conventional mass spectrometric techniques.

Ivermectin has been analyzed by mass spectrometry in other contexts. Direct injection thermospray-MS in negative ion mode has been used to confirm an LC/UV method for ivermectin in milk.[16] Thermospray-MS detection of ivermectin has also been reported by others.[17] The spectra show molecular ions and fragment ions due to loss of one or two sugar moieties, but not with adequate sensitivity for confirmation in milk. Also, direct exposure chemical ionization-MS/MS has been used to confirm ivermectin spiked into liver at the 38 ppb level.[18] In this tandem mass spectrometry approach, the first analyzer is used to filter out interferences.

On-line liquid chromatography-MS (LC/MS) is another approach to solving the problem of matrix interferences, without resorting to tandem instruments. Liquid introduction techniques for mass spectrometry offer several advantages over direct probe or gas chromatographic introduction: (1) detection of parent drug can usually be done without derivatization; (2) liquid chromatographic interfaces can be used for on-line separation prior to detection; (3) retention time can be used to aid in sample identification. Ivermectin's complex structure precludes GC analysis without extensive and undesirable pretreatment.

Experimental

The particle beam interface is a relatively new technique for on-line LC/MS. The LC eluent is nebulized by helium and passed through a two stage momentum separator to remove volatile solvents prior to entering a conventional ion source. We have found that the particle beam LC/MS interface with methane negative chemical ionization (NCI) is a very sensitive technique for detection of ivermectin. Source temperature was set to 300°C to maximize parent ion and fragment ion intensities. The specificity of this technique was based on detection of fragment ions arising from ring cleavage, loss of both sugar moieties, and one or two H_2O molecules from the

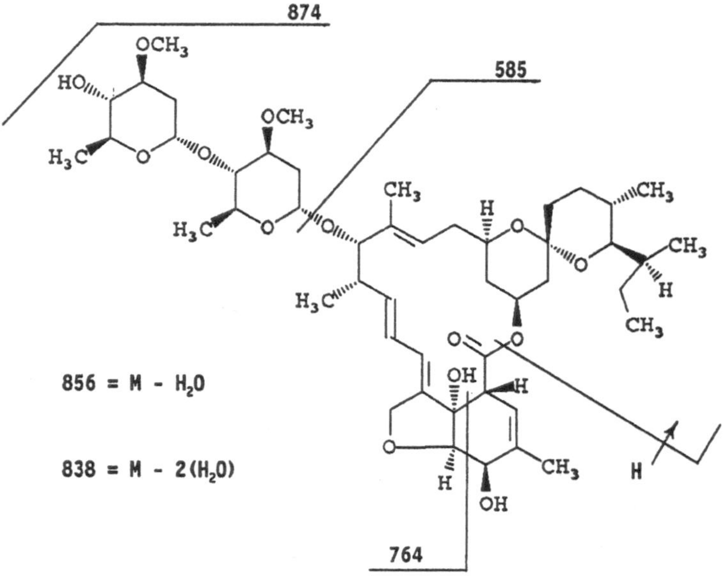

Fig. 3   Proposed structures of ivermectin H_2B_{1a} NCI fragment ions.

molecular anion, as shown in Fig. 3. Samples containing ivermectin were injected onto a Waters Nova-Pak C-18 column, 2 x 150 mm, with a mobile phase of 80% acetonitrile in pH 4 ammonium acetate buffer at a flow rate of 0.4 mL/min.

Various extraction and cleanup techniques were tested for convenience and percent recovery. For example, the liquid/liquid extraction used in the official determinative assay[19] was combined with an alumina B Sep-Pak solid phase extraction.[20] Ivermectin was extracted with the milk fat into nonpolar solvents, then back-extracted into acetonitrile from hexane. A simpler liquid/liquid extraction has been developed using acetone and hexane to separate fat from the milk. Ivermectin can also be centrifuged into the fat layer of raw, whole milk, which reduces the volume of material for further processing. Finally, the recently developed technique of MSPD was tested with milk using a C-18 bonded silica.

Results

Fig. 4 shows a mass spectrum for ivermectin in a milk sample fortified at 100 ppb, extracted by a liquid/liquid technique without further cleanup. The amount injected was equivalent to 100 ng on column. Selected ion monitoring of the molecular anion and major fragment ions from ivermectin standard further reduced detection limits to roughly 1 ng injected. When the alumina B cleanup was used with the official liquid/liquid extraction, parent and fragment ion signals from ivermectin fortified at 2 ppb in 10 mL of milk could be detected. In these experiments, 2-4 ng ivermectin were injected on column. Recovery was roughly 70%, with signal-to-noise ratio for major fragment ions greater than 10:1. Control milk did not yield signals at the retention time of ivermectin when analyzed under the same conditions.

Milk from a cow dosed with ivermectin was analyzed by the following confirmation scheme: liquid/liquid extraction of 10 mL milk, alumina B cleanup, injection of 40% of the final extract, and selected ion monitoring. Strong signals at ivermectin retention time were readily detected from milk samples previously determined to contain 4.3, 6.6, and 13.2 ppb ivermectin.

Other extraction techniques were tested on fortified control milk with less success. The fat layer from centrifuged whole milk fortified at 2 ppb could be extracted by a small scale technique, but recovery was about 45%. The MSPD technique was adversely affected by the high levels of milk fat, and gave low recovery and high detection limits.

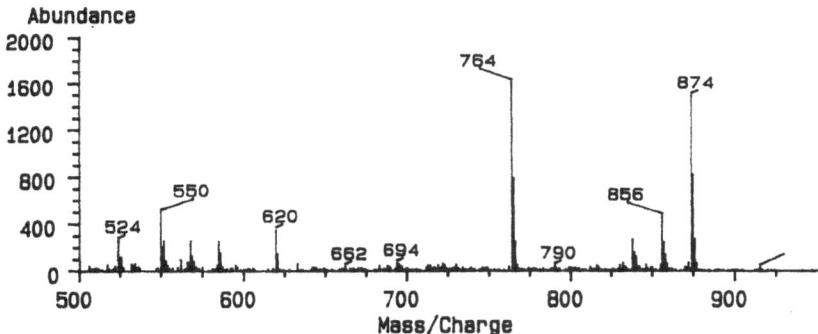

Fig. 4   Background-subtracted NCI mass spectrum of ivermectin extracted from 100 ng/mL fortified milk sample.

There are two detection advantages which result from the particle beam LC/MS approach to confirmation. First, ivermectin weighs more than the lipids remaining in the extract. Even if ivermectin is not fully resolved chromatographically, the parent ion and three of the four fragment ion chromatograms are relatively free of background signals. The LC analysis can thus be more rapid than otherwise possible. Second, material co-eluting from the column was found to increase fragment ion signals without a concurrent loss of parent ion intensity. Partial chromatographic separation produced non-interfering co-eluents which boosted the signal-to-noise on key fragment ions.

CONCLUSIONS

Milk lends itself to the use of a variety of nontraditional, rapid, and environmentally benign techniques. By using affinity chromatography, ultrafiltration, and on-column concentration,[21] tetracycline residues may determined in milk using an almost entirely aqueous based cleanup. Use of solid phase extraction reduces the amount of organic solvents needed for chloramphenicol and ivermectin extraction. The need for halogenated solvents is eliminated altogether. Finally, the development of new instrumentation, such as particle beam LC/MS, may greatly facilitate confirmation of complex residues like ivermectin. These studies show that it is possible to develop analytical methods which will detect very low levels of drug residues in milk. We hope that these procedures will soon be available for field use, helping to deter the use of illegal drugs and decrease the incidence of residues in the nation's milk supply.

ACKNOWLEDGEMENTS

The authors wish to thank Mr. H. F. Righter and the staff of DVMR's Animal Nutrition and Biology Branch for providing control milk and milk containing incurred residues. M. C. C. thanks Dr. W. A. Moats of the USDA/ARS for suggesting the use of the PLRP-S column in the TC analysis. The authors also thank Mr. M. D. Smedley and Ms. P. G. Schermerhorn for performing the second analyst checks of the TC and CAP procedures.

REFERENCES

1. *Code of Federal Regulations*, Title 21, Parts 500 to 599, Office of the Federal Register, U.S. Government Printing Office, Washington, D.C. (1990).

2. M. H. Thomas, Simultaneous determination of oxytetracycline, tetracycline, and chlortetracycline in milk by liquid chromatography, *J. Assoc. Off. Anal. Chem.* 72:564 (1989).

3. A. R. Long, L. C. Hsieh, M. S. Malbrough, C. R. Short and S. A. Barker, Matrix solid-phase dispersion (MSPD) isolation and liquid chromatographic determination of oxytetracycline, tetracycline, and chlortetracycline in milk, *J. Assoc. Off. Anal. Chem.* 73:379 (1990).

4. W. H. H. Farrington, J. Tarbin, and J. Bygrave, Trace residue analysis of tetracyclines in animal tissues, Proceedings from *Euroresidue*, p. 179 (1990).

5. W. H. H. Farrington, J. Tarbin, J. Bygrave, and G. Shearer, Analysis of trace residues of tetracyclines in animal tissues and fluids using

metal chelate affinity chromatography/HPLC, *Food Addit. Contam.* **8**:55 (1991).

6. Joint Food and Agricultural Organization/World Health Organization Expert Committee on Food Additives, Twelfth Report "Specifications for the Identity and Purity of Food Additives and Their Toxicological Evaluation: Some Antibiotics," *W.H.O Tech. Rep. Ser.* No. 430, World Health Organization, Geneva, Switzerland (1969).

7. E. H. Allen, Review of chromatographic methods for chloramphenicol residues in milk, eggs, and tissues from food-producing animals, *J. Assoc. Off. Anal. Chem.* **68**:990 (1985).

8. W. G. de Ruig, and H. Hooijerink, Determination of chloramphenicol in milk by HPLC with electrochemical detection, *Neth. Milk Dairy J.* **39**:155 (1985).

9. Chloramphenicol gas chromatographic/mass spectrometry confirmatory procedure, in: "FSIS Chemistry Laboratory Guidebook" Sec 5.023, USDA, Washington, DC (1986).

10. L. H. Keith, R. A. Libby, W. Crummett, J. K. Taylor, J. Deegan, Jr., and G. Wentler, Principles of environmental analysis, *Anal. Chem.* **55**:2210 (1983).

11. J. A. Sphon, Use of mass spectrometry for confirmation of animal drug residues, *J. Assoc. Off. Anal. Chem.* **61**:1247 (1978).

12. W. G. de Ruig, R. W. Stephany, and G. Dijkstra, Criteria for the detection of analytes in test samples, *J. Assoc. Off. Anal. Chem.* **72**:487 (1989).

13. W. J. Youden, The collaborative test, *J. Assoc. Off. Agric. Chem.* **46**:55 (1963).

14. "Compound Evaluation and Analytical Capability, National Residue Program Plan 1990", J. Brown, ed., USDA, FSIS, Washington, D. C. (1990).

15. P. L. Toutain, M. Campan, P. Galtier, and M. Alvinierie, Kinetic and insecticidal properties of ivermectin residues in the milk of dairy cows, *J. Vet. Pharmacol. Therap.* **11**:288 (1988).

16. M. Alvinerie, J. F. Sutra, P. Galtier, and P. L. Toutain, Determination of ivermectin in milk by high performance liquid chromatography, *Ann. Rech. Vet.* **18**:269 (1987).

17. D. A. Garteiz and M. L. Vestal, Thermospray LC/MS interface: Principles and applications, *LC Magazine* **3** (1985).

18. P. C. Tway, G. V. Downing, J. R. B. Slayback, G. S. Rahn, and R. K. Isenee, Confirmatory assay for ivermectin in cattle tissue using chemical ionization mass spectrometry/mass spectrometry, *Biomed. Mass Spectrom.* **11**:172 (1984).

19. F. J. Schenck, R. J. Schmid, and S. B. Clark, Determination of ivermectin in milk using high performance liquid chromatography and fluorescence detection, Food and Drug Administration *Laboratory Information Bulletin* No. 3461, Vol. 6, 1990.

20. F. J. Schenck, S. A. Barker, and A. R. Long, Rapid isolation of ivermectin from liver by matrix solid phase dispersion, *J. Assoc. Off. Anal. Chem.*, in press.

21. W. A. Moats and L. Leskinen, Determination of novobiocin residues in milk, blood, and tissues by liquid chromatography, *J. Assoc. Off. Anal. Chem.* 71:776 (1988).

THE APPLICATION OF MATRIX SOLID PHASE DISPERSION (MSPD) TO THE EXTRACTION AND SUBSEQUENT ANALYSIS OF DRUG RESIDUES IN ANIMAL TISSUES

Steven A. Barker* and Austin R. Long**

*Department of Veterinary Physiology, Pharmacology and Toxicology, Residue Studies Laboratory, School of Veterinary Medicine, Louisiana State University, Baton Rouge, LA 70803

**Animal Drugs Research Center, Food and Drug Administration Denver, CO 25087

INTRODUCTION

The contamination of our environment and of our food supply with industrial, agricultural and pharmaceutical agents possessing toxicologic and/or pharmacologic character are major man-made problems that must be redressed in the coming decade. The prevention, detection and elimination of such contamination constitute major challenges to our technology and our science and will result in significant financial resources being expended by private, State and National agencies to remediate and monitor these problems. Of particular concern is the use of veterinary drugs in the production and maintenance of food animals. The improper use of such drugs leaves residues of the parent and/or its metabolites in the tissues or milk of the animal so treated and, subsequently, in the food products derived from these sources. The presence of such drugs poses a potential threat to human health on toxicologic, immunopathological and microbiological grounds. Toxicological concerns relate to a direct toxic effect of the compound on the consumer, possibly resulting in physiological abnormalities, such as cancer. Immunopathological complications arise when the drug serves as an antigen, resulting in hypersensitivity reactions in sensitized individuals. Microbiological concerns relate to the transmittance of antibiotic resistance to bacteria from low level exposure to antibiotics and the possible emergence of resistant pathogenic bacteria in the natural flora of humans.

Attempts to gauge the nature and severity of any contamination by these agents and the assessment of any risk from their ingestion relies heavily on the development and application of sophisticated analytical technology that can rapidly isolate, identify and quantify the substances that are of concern. In this regard, the absence of such methodology for drug residues in food products of animal origin has greatly complicated and slowed such assessments. Methods have been developed for the monitoring of foods for environmental and pharmaceutical contamination by various regulatory agencies. However, existing methods for the extraction, detection and analysis of environmental and pharmaceutical residues in foods so derived tend to be highly labor and materials intensive, time consuming, and, subsequently, prohibitively expensive and lengthy to perform in large number. Such methods also tend to be too highly specific, being developed to isolate a

Analysis of Antibiotic Drug Residues in Food Products of Animal Origin
Edited by V.K. Agarwal, Plenum Press, New York, 1992

119

single compound, and, thus, require variations in methodology for each new drug, and often, each species and/or tissue examined. Indeed, a recent United States Congressional review[1] concluded that insufficient numbers of analyses for the detection of such agents were being conducted on insufficient numbers of animals and animal derived products that enter the human food chain. Presently, contamination may often go undetected or is only determined to be of concern after the animals in question have been processed and the derived products placed on the market and/or consumed. This fact is in large measure attributable to the failure of our existing approaches and methodologies to rapidly assay tissues, milk and other food products in a timely manner for the wide range of drugs that are of concern. The existence and implementation of such rapid methods would permit more products to be examined for more drugs and would afford the opportunity to prevent the inclusion of contaminated animals in the food supply by detecting them at the source.

An initial response to solving this problem has come in the growing acceptance of immunoassay (IA) and receptor tests for screening purposes. However, in many cases these tests are not considered to be adequately specific in the detection and quantitation of a particular residue and may give false-positive or difficult to interpret results. The use of such tests on milk, for example, could cause the product to be ruled as unacceptable, and be based, perhaps erroneously, on the result obtained from a false-positive from an immunoassay or receptor screening test. Such decisions are made from these preliminary IA or receptor results, however, because adequate analytical methodology does not generally exist to rapidly remove cross-reacting materials from the sample or to specifically isolate the suspect drug residue and confirm it in a timely manner. Even if such immunoassay tests can be made specific, a further complication will exist in that such tests are not generally applicable to tissues, since in order to perform these tests the drug residues must be isolated away from the tissue matrix with adequate recovery and in the absence of interferences. This fact also greatly complicates the screening of tissues by any method.

Despite these developments and problems, the increasing numbers of drugs in food animal production and the increasing numbers of drugs available, the continuing concerns over environmental pollutants and the recent decision to add aquatic species derived from aquaculture, ocean farming and fishing to foods which must be inspected, will place a significant burden on the ability of State, National and private agencies to examine more animals for more drugs and environmental contaminants in a timely fashion and, at the same time, be certain of the results obtained. However, failure to do so could pose a significant threat to human health and economic loss to the producers and their nations. Thus, it is imperative that a new philosophy and a new approach to performing such determinations be developed.[2] The advantages of immuno-or receptor assays must be underscored by the need to make such tests reliable and applicable to the drugs and matrices of concern. The approved methods developed for the purpose of isolating such drugs from tissue matrices must be rapid, simple, generic, and therefore, applicable to a range of compounds to be isolated from a given sample. Such methods are also needed to perform analyses to determine the structure and level of any drug detected and to subsequently assist in confirming its presence in as short a span of time as possible.

Toward these ends our laboratory has directed its research to the development of such an approach and such methodology. We have reported the development of a process called matrix solid phase dispersion[3] (MSPD) that addresses many of the aforementioned needs for conducting tissue residue analysis and eliminates many of the historical difficulties. This process involves the blending of a solid support (silica particles) derivatized with a lipid solubilizing polymer (C_{18}) and a sample matrix, such as tissue or milk, to produce a semi-dry, easy to handle material that can be packed into a column and sequentially eluted with various solvents. In this manner a specific drug, a class of drugs or several classes of drugs can be rapidly isolated based on their distribution between the polymer and dispersed tissue matrix and the sequence and polarity of the eluting solvents used.[2]

It is proposed that MSPD may serve as a generic technique that could speed and enhance the process of tissue analysis for a wide range of drugs by IA and/or other techniques. The results of the application of MSPD to these types of analyses are herein presented and conclusions are drawn concerning its role in changing approaches to drug residue analyses.

EXPERIMENTAL

Specific conditions for the isolation and subsequent instrumental analysis of a single drug or class of drugs by MSPD are given in the references listed for the compounds in Table 1. However, the general procedures for performing MSPD extractions are given below;

MATERIALS

a. Solvents; all solvents used should be of liquid chromatography (LC) grade or better.

b. Water; for LC analyses: distilled and passed through Modulab Polisher 1 (Continental Water Systems Corp., San Antonio, TX) water purification system.

c. Drug and Metabolite Standards; obtained from commercial sources; purity of ≥ 98%.

d. Column material; Bulk C_{18}, 40 μm, 18% load, endcapped (Analytichem International, Harbor City, CA).

e. Sample Extraction Columns; ten ml syringe barrels washed with soapy water, rinsed with distilled water and air-dried before use.

f. Filter paper discs (Whatman #1, 15 mm).

g. Syringe plunger (10 ml) with the rubber end and pointed plastic portion removed.

h. Disposable Eppendorf pipet tip (100 μl).

EXTRACTION PROCEDURE

Two grams (2 g) of C_{18} are placed in a glass or agate mortar and an aliquot (0.5g) of sample (fat, muscle, liver, kidney, milk, etc. from various sources) is placed on top of the C_{18}. Standards are injected into the sample for preparation of standard curves, recovery studies, use of internal standardization or fortification studies and the sample is allowed to stand for two minutes. Samples are then gently blended (approximately 30 seconds) into the C_{18} with a glass or agate pestle until the mixture is homogeneous in appearance. It may be necessary to scrape the sides of the mortar and pestle and repeat this process of blending for particularly wet samples or if homogeneity is not evident from the initial process. The resulting blend is transferred with a funnel to a 10 ml syringe barrel plugged with a filter paper disc (Whatman #1, 15mm). The column head is covered with an identical disc and the column contents are compressed to a volume of 4.5 ml using a syringe plunger with the rubber end and pointed plastic portion removed. A pipet tip (100μl disposable; Eppendorf) is placed on the column outlet to increase residence time of eluting solvents on the column.

In some cases tandem columns are prepared. In this configuration a quantity of florisil (1-2g), silica gel (1-2g) or normal or reversed phase packing material is added to the column prior to addition of the MSPD materials; or the eluate containing the compound(s) of interest is collected and eluted through a second column.

The resulting MSPD column may be eluted with a single solvent, a solvent series, or a solvent combination and may be eluted in the order of nonpolar -> polar or vice versa, depending on the target compound(s) to be isolated and the need for further purification prior to instrumental or immunoassay analysis. Volumes of 4-8 mls are routinely used for each solvent. Flow through the column may be controlled by gravity or by use of a vacuum manifold. The flow rate is approximately 1.0 ml/min and each solvent is removed as completely as possible before the addition of a further solvent.

The resulting eluate(s) may be evaporated, further purified if necessary and reconstituted for analysis. In many cases the eluate may be used directly for analysis, requiring no further sample processing. Solubilization of the compound(s) in less than 0.5 ml of solvent also gives a concentration factor which may enhance detection.

RESULTS

The results of the various applications of MSPD to the extraction and analysis of a wide range of drugs and contaminants from a variety of sample matrices are summarized in Table 1.

Recoveries reflect the mean and standard deviation for the respective drugs over a range of concentrations encompassing regulatory levels of concern. In many cases an assay was developed for the isolation of a single drug (neomycin[4], ivermectin[5], chloramphenicol[10], chlorsulon[13], chlorsulfuron[11,12], furazolidone[14-16], nicarbazin[17], and sulfadimethoxine[23]). In others the methods were developed to isolate a class (benzimidazoles[3,6,9], halogenated pesticides[17,18], sulfas[20-22], and tetracyclines[24,25]) or several drug classes (organophosphates[3]: benzimidazoles[3]: beta lactams[3]) from a single sample. Despite the range of drugs and matrices examined, excellent recoveries were obtainable by MSPD, whether being performed for isolation of a single drug, a drug class or several classes of drugs. The majority of results presented are for fortified samples. However, MSPD has been applied to incurred residue tissues (neomycin[4], ivermectin[5], nicarbazin[17], chlorsulon[13]) and has consistently proven equal or superior, in terms of recovery and assay variation, when compared to official methods.[4,5,13,17]

Individual sample preparation times are approximately ten minutes, on average. Solvent elution by gravity requires 8-10 minutes per solvent, but can be longer depending on the elution sequence performed. These times can be dramatically reduced by using a vacuum box for the elution. Solvent evaporation times are dependent on the solvent, quantity and the method of evaporation used. However, the solvent volume is only 4-8 ml/solvent and requires significantly less time than classical approaches that involve removal of hundreds of mls of extracting solvent.

DISCUSSION

Analytical capability is often governed by the sample preparation/extraction steps employed. In this regard, the isolation of drug residues having a broad range of polarities from a complex biological matrix such as tissue poses unique problems to the analyst. Classical methods have performed residue enrichment prior to an analysis by using

Table 1. A listing of compound classes for which MSPD extraction methodology has been established, the matrix examined, recovery obtained, (averaged over all concentrations examined) and the eluting solvent in which the compound(s) was isolated.

Compound	Matrix (ref.)	Recovery ($n \geq 20$)	MSPD Solvent
Aminoglycosides			
Neomycin	bovine kidney[4]	88.6 ± 4.6	0.1N H_2SO_4
Avermectin			
Ivermectin	bovine liver[5]	74.9 ± 7.3	CH_2Cl_2EtOAc (3:1)
Benzimidazoles			
Albendazole	milk[6,7]	81.1 ± 6.8	EtOAc
	bovine muscle[3]	73.9 ± 8.0	EtOAc
	bovine liver[8]	72.4 ± 2.6	EtOAc
	swine muscle[9]	93.0 ± 6.2	EtOAc
Fenbendazole (FBZ)	milk[6,7]	69.7 ± 8.9	EtOAc
	bovine muscle[3]	74.0 ± 11.8	EtOAc
	bovine liver[8]	62.0 ± 5.3	EtOAc
	swine muscle[9]	98.0 ± 5.3	EtOAc
FBZ-OH	milk[6,7]	94.4 ± 5.1	EtOAc
	bovine muscle[3]	68.4 ± 10.5	EtOAc
FBZ-SO_2	milk[6,7]	100 ± 4.1	EtOAc
	bovine muscle[3]	85.7 ± 15.0	EtOAc
Mebendazole	milk[6,7]	101 ± 4.1	EtOAc
	bovine muscle[3]	63 ± 4.2	EtOAc
	liver[8]	93.0 ± 5.7	EtOAc
	swine muscle[9]	85.2 ± 3.2	EtOAc
Oxfendazole	milk[6,7]	107 ± 2.3	EtOAc
	bovine muscle[3]	82.9 ± 9.5	EtOAc
	bovine liver[8]	86.8 ± 10.8	EtOAc
	swine muscle[9]	92.2 ± 7.8	EtOAc
Thiabendazole	milk[6,7]	88.7 ± 5.8	EtOAc
	bovine muscle[3]	63.8 ± 9.6	EtOAc
	bovine liver[8]	78.5 ± 1.0	EtOAc
	pork muscle[9]	85.5 ± 6.8	EtOAc
Beta Lactams			
Penicillin	bovine muscle[3]	86.3 ± 6.1	methanol
Ampicillin	bovine muscle[3]	59.8 ± 9.8	methanol

(con't)

Table 1. (con't)

Compound	Matrix (ref.)	Recovery (n ≥ 20)	MSPD Solvent
Cephalosporins			
Cephapirin	bovine muscle[3]	72.4 ± 26.5	methanol
Chloramphenicol	milk[10]	68.7 ± 8.3	EtOAc
Chlorsulfuron	milk[11,12]	94. 2 ± 5.8	CH_2Cl_2
Chlorsulon	milk[13]	99.8 ± 5.3	Et_2O
Furazolidone	swine muscle[14]	89.5 ± 8.1	CH_2Cl_2
	chicken muscle[15]	89	CH_2Cl_2
	milk[16]	81.7 ± 8.0	CH_2Cl_2
Nicarbazin	chicken liver[17]	87.8 ± 1.9	ACN
	chicken muscle[17]	84.4 ± 7.9	ACN
Organophosphates			
Coumaphos	bovine muscle[3]	76.6 ± 7.9	hexane
Crufomate	bovine muscle[3]	93.6 ± 6.4	hexane
Famfur	bovine muscle[3]	82.1 ± 8.8	hexane
Fenthion	bovine muscle[3]	85.6 ± 7.5	hexane
Pesticides			
lindane	bovine fat[18]	85 ± 3	ACN
	catfish muscle[19]	82 ± 5	ACN
heptachlor	bovine fat[18]	86 ± 5	ACN
	catfish muscle[19]	84 ± 9	ACN
aldrin	bovine fat[18]	92 ± 13	ACN
	catfish muscle[19]	94 ± 12	ACN
heptachlor expoxide	bovine fat[18]	86 ± 6	ACN
	catfish muscle[19]	93 ± 12	ACN
p,p' - DDE	bovine fat[18]	94 ± 6	ACN
	catfish muscle[19]	91 ± 6	ACN
dieldrin	bovine fat[18]	95 ± 3	ACN
	catfish muscle[19]	91 ± 2	ACN
endrin	bovine fat[18]	97 ± 3	ACN
	catfish muscle[19]	93 ± 7	ACN
p,p'-TDE	bovine fat[18]	97 ± 5	ACN
	catfish muscle[19]	97 ± 4	ACN
p,p'-DDT	bovine fat[18]	102 ± 5	ACN
	catfish muscle[19]	97 ± 5	ACN

Table 1. (con't)

Compound	Matrix (ref.)	Recovery (n ≥ 20)	MSPD Solvent
Sulfonamides (S = sulfa)			
S-diazine	swine muscle[20]	95.1 ± 15.1	CH_2Cl_2
	milk[21]	81.2 ± 4.8	CH_2Cl_2
	infant formula[22]	99.6 ± 5.3	CH_2Cl_2
S-dimethoxine	swine muscle[20]	95.8 ± 12.4	CH_2Cl_2
	milk[21]	89.6 ± 8.1	CH_2Cl_2
	infant formula[22]	103 ± 9.2	CH_2Cl_2
	catfish muscle[23]	101.1 ± 4.2	CH_2Cl_2
S-merazine	swine muscle[20]	78.1 ± 9.1	CH_2Cl_2
	milk[21]	82.0 ± 4.5	CH_2Cl_2
	infant formula[22]	92.7 ± 8.8	CH_2Cl_2
S-methazine	swine muscle[20]	84.7 ± 8.2	CH_2Cl_2
	milk[21]	92.7 ± 5.6	CH_2Cl_2
	infant formula[22]	99.1 ± 8.8	CH_2Cl_2
S-methoxazole	swine muscle[20]	95.7 ± 14.8	CH_2Cl_2
	milk[21]	89.4 ± 8.3	CH_2Cl_2
	infant formula[22]	112 ± 8.2	CH_2Cl_2
S-anilamide	swine muscle[20]	70.4 ± 12.7	CH_2Cl_2
	milk[21]	73.1 ± 7.3	CH_2Cl_2
S-thiazole	swine muscle[20]	80.3 ± 11.1	CH_2Cl_2
	milk[21]	93.7 ± 2.7	CH_2Cl_2
	infant formula[22]	75.9 ± 11.1	CH_2Cl_2
Sulfisoxazole	swine muscle[20]	92.8 ± 11.8	CH_2Cl_2
	milk[21]	88.6 ± 11.2	CH_2Cl_2
	infant formula[22]	93.1 ± 9.7	CH_2Cl_2
Tetracyclines			
Chlortetracycline	milk[24]	77.2 ± 11.3	ACN:EtOAc (3:1)
Oxytetracycline	milk[24]	93.3 ± 3.4	ACN:EtOAc (3:1)
	fish[25]	80.9 ± 6.6	ACN:EtOAc (3:1)
Tetracycline	milk[24]	63.5 ± 19.6	ACN:EtOAc (3:1)

extensive sample homogenization, solvent extractions, column chromatography and various combinations of these and other techniques. However, their utility has been limited due to the need to use large volumes (hundreds of mls) of extracting solvents, perform chemical manipulations such as pH adjustments and component precipitations, centrifugations, backwashes and solvent evaporations on a scale that prohibits their general application to a large number of samples for the analysis of a large number of drugs. Such methods have little or no utility for rapid screening techniques and could not be applied to isolate drugs from biological matrices for subsequent immunoassay (IA) or receptor analysis in any routine way. In contrast, liquid samples such as milk, urine, blood or other body fluids can often be assayed more directly. Similarly, tissues may be macerated or blended with water to prepare an aqueous extract that may be applied to such screening tests. However, it is highly unlikely that many of the drugs of concern, particularly non-polar drugs, will be present at representative levels in exudate or, necessarily, in aqueous homogenate and will require isolation and further definition of our ability to recover and determine the drugs in such matrices by IA and receptor screening tests. This raises another concern; the need to assure that, once detected, the results of such screening tests are subsequently confirmable. As has occurred in the human drugs of abuse testing field, the use of screening tests alone to make decisions to reject or to destroy product and bring into question the producer's practices or reputation is legally and morally indefensible, given the fact that false positive results from such tests will occur. This will be true whether milk, blood, urine or tissues are used to screen for contaminants. Therefore, methodology that is applicable to these matrices to isolate the drugs of concern while removing possible interferences from naturally occurring compounds and that can provide a rapid process for subsequent definition of the response from a screening test is urgently needed.

We propose that MSPD may provide some solutions to the problems so defined. The mechanisms involved in MSPD encompass sample homogenization and disruption, exhaustive extraction, fractionation and purification in a single process. The method involves the dispersal of a sample over a theoretical surface of 1,000 m^2 in a thin film (100Å), utilizing the shear forces of the particles and the blending and grinding action employed to disrupt the sample architecture while the polymer (C$_{18}$ or other) serves to dissolve and disperse the sample components on the basis of hydrophobic/hydrophilic interactions. In this manner the polymer bound to the solid support literally disrupts and unfolds cell membrane or micellar lipids, as shown by scanning electron microscopic examinations of the matrix obtained after the blending of a sample.[26,27] Examinations of tissues so treated have given no evidence of the presence of intact cells, subcellular organelles or nuclei in any sample, indicative of complete disruption. The disruption process can be envisioned as incorporating the use of shear forces from the particles with tissue solubilization using detergents, two classical approaches to tissue and cellular disruption. However, in MSPD, the "detergent" is bound to the particle and eliminates the need to subsequently remove the detergent before final analysis, and provides a unique column support for subsequent isolation of components.

Samples so treated would be expected to disperse in such a manner that the more neutral or lipophilic compounds would tend to solubilize in the polymer phase. Thus, triglycerides and the less polar ends of phospholipids, steroids and other tissue components would be expected to insert into the polymer phase that coats the surface and pores of the particles. More polar components would associate through hydrophilic interactions with themselves and with the more polar ends of the compounds inserted in the polymer. This distribution would, thus, be similar to that observed in cells themselves. Drugs and their metabolites would be expected to distribute in the same manner.

By transferring the material to a column and performing a solvent elution one obtains a distribution of the target compounds as well as other sample components that is dependent on 1) interactions with the bound polymer phase, 2) interactions with the dispersed sample matrix components, 3) molecular size and 4) interactions with the eluting solvents. This process is distinctly different from classical solid phase extraction (SPE), possessing elution and retention properties that appear to be a mix of partition, adsorption and paired ion/paired component chromatography that is somewhat unique. These properties are affected by the following variables; 1) The solid support and polymer utilized. The studies conducted to date have almost exclusively dealt with the use of 40μm silica particles having a 60Å pore size and an 18% carbon load of octadecylsilane that has been encapped. However, it would be expected that C_4, C_6 or C_8 would also be applicable as would cyclohexyl and other hydrocarbon based silanes and that mixed polarity phases such as C_8-sulphonic or true amphiphiles similar to detergents would provide unique results. For example, Schenck[4] has recently reported that the use of cyanopropyl silane derivatized silica particles gives higher recoveries and cleaner eluates than C_{18} derivatized material in the isolation of neomycin from kidney tissues. A further consideration is the particle and pore sizes utilized. Larger particles or less regular particles should also be applicable and should be less expensive. However, larger particles will have less available total surface area unless there is a proportional increase in pore size. In some cases these particles may have greater capacity per unit volume than the materials used to date. The pore size may also play a role, as in other forms of chromatography, in retaining molecules of certain sizes, giving the material a size exclusion/ inclusion character. Particles that are not end-capped may also prove useful. This will be the case where the drugs to be isolated are not adsorbed by the presence of free silicate hydroxyls or where these functional groups are overwhelmed by sample matrix effects. Similarly, silica particles alone, having no polymer, may prove useful for some applications, especially the isolation of non-polar species. However, the degree of sample matrix disruption and fractionation obtained using silica alone is not as great as that observed for polymer (C_{18}) derivatized particles.[26,27] 2) The nature of the sample matrix. We have noted that the elution profiles for specific drugs or classes of drugs are not always effected by a change in sample type, i.e. liver versus muscle versus milk. For example, regardless of the matrix the sulfa drugs are eluted from MSPD columns with methylene chloride. However, we do note differences in recovery and this may be due to differences in the distribution of sample components, such as lipids, proteins, carbohydrates and other compounds. In this regard, we have demonstrated that MSPD can also be applied to obtain quantitative fractionation of the sample components themselves, isolating lipids to DNA from the same sample by altering the elution solvent sequence[26, 27]. It is, thus, believed that one of the mechanisms involved in isolating drugs by MSPD is the association of drugs with certain sample components and their simultaneous co-elution from the column. 3) The ratio of sample to solid support. We have consistently developed methodology using a ratio of 0.5g of sample to 2.0 g of C_{18} material. However, this ratio has not been established as necessarily ideal and depending on what is to be isolated a greater or lessor ratio may suffice. This will especially be true if a greater capacity material or different polymer is used. Scale-up of the process is feasible as well, but will require examination of sample to support ratios to obtain the best result. However, large scale (5-10g of sample, 10-20g or more of solid support) MSPD columns may become too difficult to prepare and will tend to be less time and materials efficient. 4) The solvent elution sequence performed. Many of the methods developed to date have begun elution with a non-polar solvent such as hexane and have increased the solvent strength in subsequent steps, i.e. - CH_2Cl_2 -> EtOAc -> ACN -> MeOH -> H_2O, such as would be conducted for a normal phase column. Different drugs or drug classes, as well as different sample matrix components are found in each fraction. This process can thus he used to perform clean-up of the sample prior to bringing off the target compound(s) or can be used as a multi-drug, multi-drug class isolation, performing a significant degree of fractionation even if a single solvent is used. For example, the pesticides listed in Table 1 are eluted by treatment of the columns with ACN[18,19] alone. This also tends to elute some of the compounds that would otherwise have been eluted by hexane, CH_2Cl_2 and/or EtOAc. However, by not performing the previous elutions the majority of these compounds remain on the column due to the lower solvent strength of ACN for reversed phase processes. By remaining on the column the eluate is influenced by their presence and, in general, a higher recovery of the pesticides is obtained than if the previous elutions were performed. This is the case in performing elutions beginning with highly polar solvents. In this manner the column polymer and the dispersed matrix serve to retain more non-polar species, similar to mechanisms that would be observed in classical SPE on reverse phase supports.

We have observed a rather significant capability of various solvent combinations and sequences to provide fractionation of complex samples by MSPD, as implied from the results in Table 1. We have not yet examined the full use of pH gradients, or gradients per se, but believe that, for certain applications, MSPD will be applicable to high resolution fractionation of larger samples eluted with slowly changing solvent composition, particularly for the isolation of naturally occurring compounds from biological matrices. 5) Matrix Modifiers. One may influence the disruption, distribution and subsequent elution profile of an MSPD column by blending the sample in the presence of acids, bases, salts, chelators, preservatives or other modifiers. For example, the tetracyclines are obtained in higher recovery by incorporating oxalic acid and other chelating agents into the MSPD column.[24,25] Thus, one can choose to alter the polarity of the target compound by protonation or deprotonation (acid, base) so as to cause a drug or metabolite to be retained longer or eluted earlier, depending on the needs of the analyst. 6) The use of various solid support combinations or tandem column configurations. We have observed that for many of the drugs and matrices examined, little or no further clean-up or chemical manipulation of the sample is necessary following elution. However, several classes of compounds co-elute with sample matrix components that interfere with detection or that foul the instrumentation after several injections. In some cases a simple back-extraction or ne-solubilization process has eliminated such interferences. For several drugs a more efficient process has been the use of tandem columns. For example, the MSPD isolation of nine pesticides from bovine fat or catfish muscle [18,19] is assisted by including in the bottom of the same column 2 g of Florisil, which has little retention for such compounds but readily removes lipids and other materials that adversely effect subsequent GC/electron capture detection. Similarly, Schenck, et al have utilized alumina SPE columns post elution of nicarbazin from chicken liver and muscle[17] and ivermectin[5] from beef liver. We have also observed that the incorporation of up to 1 g of C_{18} in the bottom of an MSPD column prior to addition of the matrix blend can often provide extra fractionation and clean-up of eluates.[3] Thus, various configurations of columns and column packing materials, incorporated in the same MSPD column or used post-elution, could provide the capability to perform specific drug isolations and purification or be used to further enrich or enhance the level of the extracted compounds for subsequent analysis. This approach also permits the overall process to remain as time and materials efficient as possible. A further application, along these lines, would be the development and use of in-line affinity discs to remove and detect drugs of interest as they are eluted. This concept is presently being pursued on several fronts.

The primary concerns involving the use of MSPD have been directed at the fact that such a small sample size is used. The problems which arise are those of limits of detection for many compounds and the homogeneity or representative character of the sample. We have addressed the latter issue by suggesting that a larger sample of 5, 10 or 50g be pre-blended by mechanical means, much as is presently done for many residue analysis methods, and that sub-samples of this blend be utilized for MSPD based methods. This method has been applied to a number of drugs assayed from incurred residue tissues and has shown that MSPD can give highly consistent and reproducible results. In most cases the assay variation for MSPD has proven to be less than that observed for official or approved methods.[4,5,13,17] The issue of sample size and limits of detection is more troubling to some. However, the limits of detection of many assays are adversely affected by interfering co-extractants. The level of such interferences can often be proportional to the size of the sample used. A smaller sample does reduce the total quantity of the drug to be isolated but likewise decreases the interferences. Another variable in this argument is the final volume of the extract. A 10g sample containing $0.1 \mu g/g$ of a target drug extracted and dissolved in a final volume of 1 ml has the same concentration of drug/ml as the use of 0.5g of tissue extracted and dissolved in $50 \mu l$ of solvent. However, the 10g sample will require more manipulation, solvent and technician time to prepare than the corresponding smaller sample and will also require greater efficiency in the separation and detection technology utilized in its analysis. The small sample size for MSPD will limit its use, however, in cases where extremely low limits of detection are required (1-5ng/g) or when particularly sensitive methods of detection are not available or applicable. MSPD will not work in such cases. However, this should not be a complication in the possible use of MSPD for the isolation of drugs from tissues followed by IA analysis, however, and will, of course, depend on the sensitivity and configuration of the immuno- or receptor assay performed. One can combine extracts from several MSPD columns prepared from the same tissues to enhance detection. In some cases, this may still be more practical than classical methods but should not be considered as a general solution.

We have recently found that one may use MSPD to perform extractions even more rapidly and simply than previously described. All methods developed to date rely on the elution of drugs from the blended tissue matrix packed in a column. We have found that for drugs such as beta lactams and aminoglycosides that one may perform MSPD on 100mg of tissue using 400mg of C_{18}, placing the blend in a tube with 1.0mL of 0.01M HCl, briefly sonicating (15sec) and, after centrifugation (5 min), obtain the compounds with recoveries of greater than 70% in the supernatant. This supernatant is suitable for immunoassay analysis after combining with assay buffer. Other drugs can be isolated using different solvents such as methylene chloride for sulfa drugs, with the solvent being removed by evaporation and replaced with assay buffer prior to analysis. This approach is accomplished through the same mechanisms as described for MSPD conducted in a column; the C_{18} retains more lipophilic compounds and the unique C_{18}/tissue phase produced releases compounds as a function of their solubility in the eluting solvent and this phase. This microscale approach may be best applied to the rapid isolation of drugs from tissues for the subsequent application of immunoassay or other methods of screening. We believe that this approach will have definite advantages over the simple maceration of tissue, using the exudate to assay for drugs, or the need to mechanically homogenize and blend the sample in the presence of solvents for subsequent screening.

Thus, MSPD may prove applicable on several levels for residue monitoring programs; rapid isolations or preparation steps for samples prior to immuno- or receptor assay screening, extraction for determination methods and final preparation of samples for subsequent confirmation. MSPD greatly reduces the sample manipulation, solvent usage and disposal, technician time and the subsequent time to obtain results for a given analysis. It is also amenable to automation. MSPD is a generic approach for the disruption and extraction of tissues and may be applied to the isolation of a wide range of drug residues, environmental contaminants and naturally occurring compounds from a single biological matrix. The speed, simplicity and broad capabilities of this approach should lend itself to the analysis of tissues, milk, blood, etc. on site, where required, particularly when coupled with appropriate immunoassay formats. The generic nature of MSPD may also provide more rapid method development and allow us to keep pace with the ever increasing number of drugs that must be monitored.

A new approach and philosophy needs to be implemented for the better regulation of residues in animals destined to become part of the human food chain. This must include prevention and detection at the source. Since tissues are the most difficult to assay a greater use must be made of blood and urine screening. In many cases correlations can be drawn concerning possible violations in tissues and the levels of drug detectable in these body fluids. Greater availability and use of simple tissue residue isolation procedures and immunoassays will also make such an approach feasible at this level. However, these developments must be supported by an advanced capability to quantitate and confirm suspect product prior to its rejection.

We recognize that MSPD will not be applicable to all of these and other problems to be encountered in conducting tissue residue analysis. However, we do believe that it may serve as a major adjunct to such analytical procedures, either as presently defined or as it may be modified in the near future and that it offers an enhanced capability to perform the most difficult aspect of drug residue analysis; the isolation of the target compounds from the sample matrix.

Acknowledgements

The studies presented were supported by grants and cooperative agreements FD-U-000235, 5V01-FD-01319 and FD-U-000581 with the Food and Drug Administration, Center for Veterinary Medicine and the Egyptian Peace Fellowship Program, PF #3497, United States Agency for International Development. The work of Mr. Frank J. Schenk, Dr. Murray E. Hines, II, Ms. Lily Hsieh, Ms. Marsha Malbrough and Dr. Charles R. Short are expressly recognized as is the assistance of Ms. Wendy Looney, Tanya Richmond and Fatemeh Moharrernezhad.

References

1. Union Calendar #274 House Report 99-461 "Human Food Safety and the Regulation of Animal Drugs" 27th Report by the Committee on Government Operations. 115 pg. December 31, 1985.

2. Barker, S.A., Long, A.R., and Short, C.R., 1989. A new approach to the isolation of drug residues from animal tissues. Proceedings of the Sixth Symposium on Vet. Pharmacology and Therapeutics. W. Huber, ed., American Academy of Veterinary Pharmacology and Therapeutics. Blacksburg, Virginia. June, 1988. p. 55-56.

3. Barker, S.A., Long, A.R. and Short, C.R., 1989. Isolation of drug residues from tissues by solid phase dispersion. J. Chromatogr., 475:353-361.

4. Schenck, F.J., 1991. Matrix solid phase dispersion extraction and liquid chromatographic determination of neomycin in bovine kidney tissue. Laboratory Information Bulletin 7:3559.

5. Schenck, F.J., Barker, S.A. and Long, A.R., 1989. Rapid isolation of ivermectin from liver by matrix solid phase dispersion. Laboratory Information Bulletin, 5:3381.

6. Long, A.R., Hsieh, L.C., Short, C.R. and Barker, S.A., 1989. Isocratic separation of seven benzimidazole anthelmintics by high performance liquid chromatography with photodiode array characterization. J. Chromatogr., 475:404-411.

7. Long, A.R., Hsieh, L.C., Malbrough, M.S., Short, C.R. and Barker, S.A., 1989. Multiresidue method for isolation and liquid chromatographic determination of seven benzimidazole anthelmintics in milk. J. Assoc. Off. Analyt. Chemists. 72: 739-741.

8. Long, A.R., Malbrough, M.S. Hsieh, L.C., Short, C.R., and Barker, S.A. 1990. Matrix solid phase dispersion (MSPD) and liquid chromatographic determination of five benzimidazole anthelmintics in beef liver tissue. J. Assoc. Off. Analyt. Chemists. 73:860-863.

9. Long, A.R., Hsieh, L.C., Malbrough, M.S., Short, C.R. and Barker, S.A. 1990. Matrix solid phase dispersion (MSPD) isolation and liquid chromatographic determination of five benzimidazole anthelmintics in pork muscle tissue. J. Food Comp. Analysis 3:20-26.

10. Long, A.R., Hsieh, L.C., Bello, A.C., Malbrough, M.S., Short, C.R. and Barker, S.A., 1990. Method for the isolation and liquid chromatographic determination of chloramphenicol in milk. J. Agric. Food Chem. 38:427-429.

11. Long, A.R., Hsieh, L.C., Malbrough, M.S., Short, C.R. and Barker, S.A. 1989. Method for the isolation and liquid chromatographic determination of chlorsulfuron in milk. Laboratory Information Bulletin, 5:3364.

12. Long, A.R., Hsieh, L.C., Malbrough, M.S., Short, C.R. and Barker, S.A., 1989. Isolation and gas chromatographic determination of chlorsulfuron in milk. J. Assoc. Off. Analyt. Chemists. 72:813-815.

13. Schenck, F.J., Barker, S.A. and Long A.R., 1990. Rapid determination of clorsulon in milk extracted using matrix solid phase dispersion. Laboratory Information Bulletin, 6:3426.

14. Long, A.R., Hsieh, L.C., Malbrough, M.S., Short, C.R. and Barker, S.A. 1990. Matrix solid phase dispersion (MSPD) isolation and liquid chromatographic determination of furazolidone in pork muscle tissue. J. Assoc. Off. Analyt. Chemists. 74:292-294.

15. Soliman, M.M., Long, A.R. and Barker, S.A. 1990. Matrix solid phase dispersion (MSPD) isolation and liquid chromatographic determination of furazolidone in chicken muscle tissue. J. Liquid Chromatogr. 13:3327-3337.

16. Long, A.R., Hsieh, L.C., Malbrough, M.S., Short, C.R. and Barker, S.A., 1990. Method for the isolation and liquid chromatographic determination of furazolidone in milk. J. Agric. Food Chem. 38:430-432.

17.Schenck, F.J., Barker, S.A., Long, R. and Matusik, J. 1989. Extraction and determination of nicarbazin drug residues from tissue by solid phase dispersion. Laboratory Information Bulletin. 5:3354.

18.Long, A.R., Soliman, M.M., and Barker, S.A. 1991. Matrix solid phase dispersion (MSPD) extraction and gas chromatographic screening of nine chlorinated pesticides in beef fat. J. Assoc. Off. Anal. Chem., 74:493-496.

19.Long, A.R., Hsieh, L.C., Short, C.R., and Barker, S.A. 1991. Multiresidue matrix solid phase dispersion (MSPD) extraction and gas chromatographic screening of nine chlorinated pesticides in catfish (Ictalurus punctatus) muscle tissue. J. Assoc. Off. Anal. Chemists. 74:667-670.

20.Long, A.R., Hsieh, L.C., Malbrough, M.S., Short, C.R. and Barker S.A. 1990. Multiresidue method for the determination of sulfonamides in pork tissue. J. Agric. Food Chem. 38:423-426.

21.Long, A.R., Short, C.R. and Barker, S.A. 1990. Multiresidue method for the isolation and liquid chromatographic determination of sulfonamides in milk. J. Chromatogr. 502:87-94.

22.Long, A.R., Hsieh, L.C., Malbrough, M.S., Short, C.R. and Barker, S.A. 1989. A multiresidue method for the isolation and liquid chromatographic determination of seven sulfonamides in infant formula. J. Liquid Chromatogr., 12:1601-1612.

23.Long, A.R., Hsieh, L.C., Malbrough, M.S., Short, C.R., and Barker, S.A., 1990. Matrix solid phase dispersion (MSPD) isolation and liquid chromatographic determination of sulfadimethoxine in catfish (Ictalurus punctatus) muscle tissue. J. Assoc. Off. Anal. Chem. 73:868-871.

24.Long, A.R., Hsieh, L.C., Malbrough, M.S., Short, C.R. and Barker, S.A., 1990. Matrix solid phase dispersion (MSPD) isolation and liquid chromatographic determination of oxytetracycline, tetracycline and chlortetracycline in milk. J. Assoc. Off. Analyt. Chemists. 73:379-384.

25.Long, A.R., Hsieh, L.C., Malbrough, M.S. Short, C.R. and Barker, S.A. 1990. Matrix Solid Phase Dispersion (MSPD) isolation and liquid chromatographic determination of oxytetracycline in catfish (Ictalurus punctatus) muscle tissue. J. Assoc. Off. Analyt. Chemists. 73:864-867.

26.Barker, S.A. and Long, A.R. 1990. Matrix solid phase dispersion (MSPD): A new isolation technique for chemical residues and natural products. Abstract No. 859, 41st Pittsburgh Conference and Exposition on Analytical Chemistry and Applied Spectroscopy, New York, NY.

27.Hines, M.E., Long, A.R., Snider, T.G. and Barker, S.A. 1991. Lysis and fractionation of Mycobacterium paratuberculosis and Escherichia coli by matrix solid phase dispersion, Anal. Biochem. 195:197-206.

LIQUID CHROMATOGRAPHIC APPROACHES TO DETERMINATION OF B-LACTAM ANTIBIOTIC RESIDUES IN MILK AND TISSUES

William A. Moats

Meat Science Research Laboratory
Beltsville Agricultural Research Center
Agricultural Research Service, USDA
Beltsville, MD 20705-2350

Determination of antibiotic residues including B-lactams in milk and tissues has traditionally been done by microbiological assays. Microbiological assays can detect very low levels of B-lactam antibiotics in milk and are thus well suited as screening procedures for residues. They can not, however, distinguish B-lactam antibiotics from one another. Several sensitive rapid screening tests have also been described for detection of antibiotic residues including B-lactams in milk.[1,2] These include enzyme inhibition (Penzym), a receptor assay (The Charm test), and various immunoassays. The reported sensitivities of several types of screening procedures for B-lactam antibiotics in milk are summarized in table I. These are capable of detecting several B-lactam antibiotics at levels of <10 ppb in milk. The Charm test is reported to detect as little as 2 ppb of penicillin G and cephapirin. All are less sensitive for cloxacillin with the lowest reported detection limit being about 20 ppb.

Table 1.   Detection Limits of some B-lactam Antibiotics
by Screening Procedures[a]

| Antibiotic | Test sensitivity (ng/ml) | | | |
	Disc Assays	Spot Test	Penzym	Charm II
Penicillin G	3	5	6	2
Ampicillin	5	ND[b]	10	5
Cephapirin	8	4	11	2
Cloxacillin	35	20	150	35

[a]Adapted from Hsu, et al.[1]

[b]ND — Not detectable

Analysis of Antibiotic Drug Residues in Food Products of Animal Origin
Edited by V.K. Agarwal, Plenum Press, New York, 1992

133

Specific confirmatory tests are needed to identify and quantitate residues detected by screening tests. These must equal or exceed the sensitivity of screening tests. They should also meet the requirements of regulatory agencies which vary from country to country. For penicillin G, these can range from 0-50 ppb in tissue (U.S.) and 10 ppb in milk (U.S.)[3] to 4 ppb in both in the European Economic Community and 3 ppb for both in Germany.[4] In the United States, penicillin G, cloxacillin, ampicillin, amoxicillin, and cephapirin are the principal B-lactams used in veterinary practice and therefore, of concern as residues in animals. Additional B-lactams may be used in other countries.

Some type of chromatographic separation is necessary to identify individual antibiotics. B-lactams are fairly polar, nonvolatile, and somewhat heat sensitive. Of common chromatographic modes, high performance liquid chromatography would seem to be the method of choice. However, gas-liquid chromatography (GLC) has been successfully used for determination of B-lactam antibiotics with neutral side-chains.[5] GLC requires lengthy sample preparation and derivatization and cannot be used with amphoteric B-lactams. Thin layer chromatography (TLC) can be an inexpensive alternative and has been used with colorimetric detection[6] or detection by bioautography.[7,8,9] Supercritical fluid chromatography (SFC) is not suitable for determination of B-lactams.

We have investigated several types of LC columns for determination of penicillins. Reversed-phase columns were most satisfactory for residue analysis. They gave good separation from interferences and retention times were not affected by other substances in sample extracts. Ion-exchange columns did not give good separation from interferences and retentions were greatly affected by other substances in sample extracts. Polymeric styrene divinylbenzene copolymer columns were equivalent to C18 bonded columns and had slightly different selectivity which was advantageous in some cases. Reported HPLC methods for determination of B-lactam antibiotic residues in milk and tissues are summarized in table 2. Development of methods of the needed sensitivity has proven elusive and it is only recently that methods with detection limits of <10 ppb have been described.[4-6,11,21-24] A variety of approaches have been used. Methods used for extraction/deproteinization include:

1. Deproteinization with organic solvents (methanol or acetonitrile)[4,5,10-12,15,20-24]
2. Deproteinization with tungstic acid[14,25]
3. Ultrafiltration (of milk)[16-18]
4. Direct solid-phase extraction (of milk)[13,19]

Except for extracts prepared by ultrafiltration, some additional concentration and cleanup was used.

B-lactams have been determined either directly by UV absorption or after derivatization. A number of investigators have used derivatization with triazole,[23,25] or hydrazine[10] and mercuric chloride to form products which absorbed at a longer wavelength or which fluoresced. Wiese and Martin[26] observed that the molar absorbance of penicillin G at 193 nm (38,000) was greater than that of the triazole mercuric chloride derivative at 325 nm (20,000). Derivatization in this case moved the absorbance maximum to a region where fewer interferences absorbed but did not increase sensitivity. Berger and Petz[4] used derivatization with 4-bromomethyl-7-methoxycoumarin to form an ester which was determined fluorometrically. Precolumn derivatization changes both the spectral and chromatographic properties of the compound which may or may not be advantageous. On-line post column derivatization of compounds with free amino groups such as ampicillin and amoxicillin is also possible but has not been used for residue analysis. Precolumn derivatization is generally time-consuming and can introduce unwanted reaction products. Our laboratory has therefore based methods on detection by UV absorption. This is simple and effective but requires efficient chromatographic methods of separating compounds of interest from interferences. The retentions of B-lactams on reversed-phase columns differ so greatly that only a few compounds can be determined in a single isocratic analysis. All compounds can be determined in a single gradient elution procedure.[11,12] However, separation from interferences was less satisfactory with gradient elution. Isocratic analysis does not always clear the column of interferences before the next sample is injected. We have sometimes found it necessary to include a flush after each sample.[22]

In our laboratory, we have attempted to develop efficient methods of separating B-lactams from interferences so that they can be determined directly by UV absorption, thus eliminating the need for lengthy derivatization procedures. Extraction/deproteinization by addition of acetonitrile to milk or tissue homogenates is simple, rapid, and reproducible. An aliquot of filtrate was taken as representative of the sample-acetonitrile mixture so there was no need to wash the precipitate. This method gave 90-100% recoveries of the monobasic penicillins, penicillin G, penicillin V, and cloxacillin.[11,12,20-22] However, recoveries of amphoteric B-lactams were not as good.[24] We are currently working on improving recoveries of these compounds. Methanol was less effective for deproteinization. Methanol should not be used for sample preparation or analysis of B-lactams since these compounds can undergo methanolysis.[27]

Table 2. HPLC Methods for Determination of B-lactam Antibiotic Residues in Foods

Substrate	Compounds	Extraction/Cleanup	Detection	Sensitivity (ppb)	Reference
Milk	8 with neutral side-chains	CH$_3$CN/partitioning	Fluorimetric after derivatization	20–55	10
Milk	PG,PV,Clox[a]	CH$_3$CN/partitioning	UV 220	5PG,PV 2 Clox	11
Tissue Beef, Pork	PG,Clox	CH$_3$CN/partitioning	UV 220	50PG 20 clox	12
Milk	PG,PV,Amp	Solid-phase C18/ Solid-phase	UV 220	30	13
Tissue	PG	Tungstic acid/ Sep-pak, alumina	UV 210	50	14
Fish	Amp	Methanol/Florisil	UV 222	30	15
Milk	PG	Ultrafiltration	UV 210 LS-MS	10 100	16
Milk	Amp,Amox, Clox	Ultrafiltration	UV 210	10 Clox, 25 Amp 50 Amox	17
			LC-MS	25 Clox, 50 Amp 25 Amox	

Continued on next page

136

Continued from previous page

Matrix	Antibiotic	Extraction	Detection	LOD	Ref.
Milk	Cep	Ultrafiltration	UV 291 LC-MS	10 100	18
Milk	Cep	Solid-phase/ solid-phase	UV 254	10	19
Milk	Pen G,	CH₃CN/HPLC	UV 210	2	20
Milk	Pen V, Clox	CH₃CN/HPLC	UV 210	1-2	21
Tissue (Beef, Pork)	Pen G, Pen V Clox	CH₃CN/partitioning	UV 210	5	22
Milk	8 with neutral side-chains	CH₃CN/partitioning	Derivatization fluorimetric	1	4
Milk	PG	CH₃CN/partitioning	Derivatization UV 325	0.5	23
Milk	Amp	CH₃CN/HPLC	UV 210	5	24
Tissue	PG, PV	Tungstic acid/ solid-phase	Derivatization UV 325	5	25

[a]PG = penicillin G, PV = penicillin V, Clox = cloxacillin, Amp = ampicillin, Amox = amoxicillin, Cep = cephapirin

Further concentration and cleanup of the acetonitrile filtrates was required prior to analysis. Some approaches used successfully were:

1. Direct evaporation to a small volume. Turbidity was removed by washing with a small volume of hexane and acetonitrile in a centrifuge tube or by direct filtration.[20,21,24]

2. Addition of hexane and methylene chloride (as a solvent bridge) to the filtrate. The water layer was collected and further evaporated. This removed most of the coextracted lipid but required more organic solvent. Procedures 1 and 2 worked with extracts from milk but not with extracts from tissues.[20,21,24]

3. Partitioning cleanup. Acid and methylene chloride were added to the filtrate converting monobasic penicillins to the acid forms which were soluble in methylene chloride. After adding hexane, the penicillins were then extracted back into a small volume of pH 7 buffer.[11,12,22]

In extracts of tissue, the monobasic penicillin anions were bound to some substance in the filtrate, probably phospholipids which are cationic. The acid extraction was required to break this complex. The partitioning gave sufficient cleanup for analysis of tissues for penicillin V, penicillin G, and cloxacillin.[22] The partitioning cleanup could not be used with amphoteric compounds, however.

Extracts prepared by procedures 1 and 2 required further cleanup prior to analysis. Use of solid-phase extraction on disposable cartridges or laboratory packed cleanup columns was not reproducible and did not in any case give adequate cleanup. We therefore concluded that it would be necessary in some cases to use the HPLC system for both cleanup and analysis to achieve the desired sensitivity. When the HPLC system was used for cleanup, conditions were selected for cleanup and analysis which would give large differences in retention time. The penicillins would thus be separated from interferences in the initial fractions collected.

On-line concentration and gradient elution were used. The analytes eluted as narrow bands, usually in less than 0.5 ml.[20,21] Use of the liquid chromatograph for cleanup may at first glance seem complicated compared with solid-phase extraction on disposable cartridges or partitioning. However, in practice it can be fully automated so that the only operator manipulations required are loading the autosampler and adjusting the volume of the fractions collected prior to subsequent analysis. Results are reproducible since the same column is used repeatedly. The column is automatically flushed and regenerated after each run.

The requirements for an HPLC cleanup system are:

1. An autosampler capable of loading and injecting 2 ml or more of sample extract.

2. A programable pump system capable of generating a ternary gradient.

3. A fraction collector capable of collecting time windows.

4. A data system.

While it is convenient to use separate systems for cleanup and for analysis, a single system can be used for both.

A typical cleanup sequence used is as follows:

1. 2 ml of sample extract was loaded into a 2 ml autosampler loop

2. The sample was loaded onto the HPLC column directly in 100% aqueous buffer(A)

3. After 3 minutes an acetonitrile(B) gradient was started according to the program 100A:0B(0-3min) - 40:60(25-30min) - 100A:0B-(31-40 min). At 40 min, loading of the next sample is begun.[20,21,24]

A single sample can be ready for subsequent analysis in 20 min. Since cleanup is done sequentially, about 45 min additional is required for each sample.

There are a number of methods of changing both the absolute retentions and the retentions of compounds relative to one another. These include:

1. Changing the concentration and/or type of organic modifier. Changing concentrations usually does not significantly affect relative retentions. With UV detection at 210 nm, acetonitrile is the only suitable organic modifier.

2. Changing the type of column packing. For example, the selectivities of the polymeric and the bonded C18 columns were somewhat different and they could not be used interchangeably.[20-22]

3. Ion-suppression by manipulating the pH, ion pairing by adding anions or cations attached to alkyl chains or combinations of the two, are very effective in changing relative and absolute retention tissues.

4. Crown ethers increase retention of many compounds.[28] All of these procedures can change retentions of both analytes and interferences. Crown ethers were less effective in separating penicillins from interferences than ion-pairing and/or in suppression.

The effects of pH and ion-pairs on retention of five B-lactams on two types of reversed-phase columns are summarized in Table 3. Because retentions differed greatly under isocratic conditions, an identical

Table 3. The Effect of pH and Ion-Pairs on Retention of Some B-Lactam Antibiotics[a]

		Retention Time (Min)								
		Penicillin G		Cloxacillin		Ampicillin		Cephapirin		Amoxicillin	
Buffer	ion pair	PIRP-S[b]	LC-18	PIRP-S	LC-18	PIRP-S	LC-18	PIRP-S	LC-18	PIRP-S	LC-18
0.01M pH 7 phosphate	None	18.97	19.80	21.74	22.09	16.66	17.47	17.21	17.47	14.20	14.55
	0.005M Et₄NCl[c]	19.47	20.15	22.11	22.42	17.35	17.77	17.76	17.77	14.97	14.90
	0.005M SHS[d]	18.97	17.13	20.98	21.20	14.88	15.40	15.35	14.72	9.93	12.23
0.01M NH₄H₂PO₄ (pH 4.6)	None	19.30	19.98	22.07	22.60	14.72	15.58	16.98	16.99	7.71	12.41
	.005M Et₄NCl	19.86	20.69	22.68	23.43	14.73	15.79	17.00	17.08	13.20	12.83
	.005M SHS	20.78	21.42	24.29	25.09	17.17	18.69	16.40	16.90	15.93	14.11
0.01M H₃PO₄ (pH 1.6)	None	24.53	25.25	28.21	28.55	16.77	19.24	16.34	17.14	14.60	15.96
	.005M Et₄NCl	24.52	25.21	28.21	28.51	16.54	18.18	16.17	16.34	14.01	14.94
	.005M SHS	21.48	21.47	28.30	28.39	19.80	21.47	18.93	20.02	18.32	19.67

[a]Gradient elution: Buffer(A)-Acetonitrile(B) 100A:0B(0-1min)-40A:60B(25min)
[b]PIRP-S - Polymer Laboratories PIRP-S styrenedivinylbenzene copolymer (5um)
LC-18 - Supelco LC-18-DB (5um)
[c]Et₄NCl - Tetraethyl ammonium chloride
[d]SHS - sodium heptane sulfonate

solvent gradient was used with each mobile phase to facilitate comparison. An alkyl sulfonate, sodium heptane sulfonate was used as an anion-pair and tetramethyl ammonium chloride was used as the cation pair. Retentions were also affected by the length of the alkyl chain on the ion-pair but these are not shown in the table. The retention patterns were similar on the bonded silica and the polymeric column. The polymeric column was slightly less retentive. Under isocratic conditions, retention on the polymeric column was the same as that on bonded C18 columns when the acetonitrile concentration was 4% less. Retention of the penicillins with neutral side-chains, cloxacillin and penicillin G was markedly increased at acid pH as they were converted from the salt to the unionized acid form (ion-suppression). Retentions of the amphoteric compounds were similar at pH 7 and 1.6 but were markedly reduced at pH 4.6, near the isoelectric points. The counterions decreased retention of penicillins when they were of the same charge and increased retention when they were of the opposite charge. Thus, the alkyl sulfonates decreased retention of all compounds at pH 7 where all compounds were predominantly negatively charged and markedly increased retention of the amphoteric compounds at pH 1.6 since they are present as cations at that pH. Tetraethyl ammonium chloride had the opposite effect with the amphoterics. Retention of penicillin G and cloxacillin was unaffected by counterions at acid pH since the compounds were not ionized. This information provides a guide as to how retentions of the various compounds might be manipulated using combinations of pH and counterions. Manipulation of pH and ion-pairs and column type also affects the retention of interferences in a manner which is not always predictable.

The HPLC cleanup approach has been successfully applied to determination of penicillin G,[20] cloxacillin, and penicillin V,[21] and ampicillin[24] in milk. A different mobile phase was required for subsequent isocratic analysis of each penicillin because retentions differed so greatly under isocratic conditions. From stability considerations, cleanup at pH 7 was preferred. Analysis at acid pH in 0.01M H_3PO_4 shifted the retentions of the monobasic penicillins (penicillin G, penicillin V, cloxacillin) well away from interferences in the fractions. However, separation of penicillin G from some minor interferences was improved at pH 2.0 showing that further fine tuning of conditions was sometimes beneficial. These penicillins were readily quantitated at 10 ppb with detection limits of 1-2 ppb and recoveries of 90% or better. In this procedure, the polymeric styrene divinylbenzene was used both for cleanup and analysis. The unique selectivity of the polymeric column was essential for effective separation from interferences.

The approach was also used for determination of ampicillin in milk.[24] Recoveries were only 60-70% using deproteinization with acetonitrile. The retention of ampicillin is about the same at pH 1.6 and pH 7. However, addition of an alkyl sulfonate at acid pH markedly increased retention and was the basis of the separation. Separation from interferences was not as good as with the monobasic penicillins. We are doing further work on the procedures for amphoteric B-lactams.

A method was also developed for residues of penicillin G, penicillin V, and cloxacillin in tissue using a partitioning cleanup.[22] As I mentioned previously, we found that these penicillins were complexed with something, probably phospholipids, in acetonitrile extracts. This complex was broken by partitioning into methylene chloride at acid pH. The penicillins were then partitioned back into a small volume of pH 7 buffer. This gave adequate cleanup for isocratic analysis. Again, the unique selectivity of the polymeric column was necessary for efficient separation of the penicillins from interferences. Analysis was done in pH 7 buffer. Residues were confirmed by running a duplicate sample after treatment with penicillinase.

Absence of a chromatographic peak with the proper retention time provides unequivocal evidence that a given residue is not present above the detection limits of the method. Treatment of a duplicate sample with penicillinase provides a simple and sensitive confirmatory test for the B-lactams.[13,14,20-24] The blank thus obtained can be subtracted from the untreated sample which improves sensitivity and accuracy for determination of low levels of penicillin. Wiese and Martin[23] described a "digital subtraction" technique in which the penicillinase blank was subtracted electrically from the sample. This gave a sensitivity of well under 1 ppb for penicillin G in milk. Tissues required to destroy penicillins with penicillinase range from a few minutes with penicillin G to several hours with some others.[21,24]

Methods using mass spectrometry interfaced with liquid chromatography (LC-MS) have also been described for confirmation of B-lactams. LC-MS requires expensive equipment and reported methods are less sensitive than UV detection.[16-17] Diode array spectrophotometry can be used. The reliability of this method falls off at low-levels since small amounts of interferences can distort spectra.

In summary, HPLC is generally the method of choice for determination of B-lactam antibiotics in milk and tissues. HPLC is suitable for determination of all B-lactam antibiotics and generally requires less sample preparation than other modes. Some approaches developed in our own and other laboratories have been described which achieve required sensitivities for B-lactams with neutral side-chains. No entirely satisfactory methods have been reported for determination of amphoteric B-lactam antibiotics. The HPLC provides a simple and effective cleanup system in situations where rigorous cleanup is required. Further confirmation of B-lactams by treatment with B-lactamase is simple, inexpensive and sensitive.

ACKNOWLEDGMENT

The author thanks Miau Huang for technical assistance in determining the effects of buffers and ion-pairs on retention times.

REFERENCES

1. H. Y. Hsu, F. F. Jewett, Jr., and S. E. Charm. What is Killing the Bugs in Your Starter Culture? Cult. Dairy Prod. J. 22:18 (1987).

2. G. F. Senyk, J. H. Davidson, M. M. Brown, E. R. Hallstead and J. W. Sherban. Comparison of Rapid Tests used to Detect Antibiotic Residues in Milk. J. Food Prot. 53:158 (1990).

3. Code of Federal Regulations Title 21:556.510. U.S. Government Printing Office, Washington, DC (1989).

4. K. Berger and M. Petz. Labeling of penicillin residues with 4-bromoethyl-7-methoxycoumarin for HPLC with fluorimetric detection in "Residues of Veterinary Drugs in Foods", N. Haagsma, A. Ruiter, and P. B. Czedik-Eysenberg, Ed, Noordwijkerhout, The Netherlands, 1990, p. 118.

5. V. Meetschen and M. Petz. Sensitive confirmatory method for determination of seven penicillins in bovine tissues and milk. J. Assoc. Off. Anal. Chem. 73:373 (1990).

6. W. A. Moats. Detection and semiquantitative estimation of penicillin G and cloxacillin in milk by thin-layer chromatography. J. Agric. Food Chem. 31:1348 (1983).

7. C. D. C. Salisbury, C. E. Rigby, and W. Chan. Determination of antibiotic residues in Canadian Slaughter animals by thin-layer chromatography-bioautography. J. Agric. Food Chem. 37:80 (1989)

8. D. V. Herbst. Identification and determination of four B-lactam antibiotics in milk. J. Food Prot. 45:450 (1982).

9. H. Yoshimura, O. Itoh, and S. Yonezowa. Microbiological and thin layer chromatographic identification of benzylpenicillin and ampicillin in animal body. Jpn. J. Vet. Sci. 43:833 (1981).

10. R. K. Munns, W. Shimoda, J. E. Roybal and C. Vieira. A multi-residue method for the determination of eight neutral B-lactam penicillins in milk fluorescence - HPLC. J. Assoc. Off. Anal. Chem. 68:968 (1985).

11. W. A. Moats. Determination of penicillin G, penicillin V, and cloxacillin in milk by reversed-phase high-performance liquid chromatography. J. Agr. Food Chem. 31:880 (1983).

12. W. A. Moats. Determination of penicillin G and cloxacillin residues in beef and pork tissues by high-performance liquid chromatography. J. Chromatogr. 317:311 (1984).

13. H. Terada and Y. Sakabe. Studies on residual antibacterial in foods. IV. Simultaneous determination of penicillin G, penicillin V, and ampicillin in milk by high-performance liquid chromatography. J. Chromatogr. 348:379 (1985).

14. H. Terada, M. Asanoma, and Y. Sakabe. Studies on residual antibacterial in foods. III. High-performance liquid chromatographic determination of penicillin G in animal tissues using an on-line pre-column concentration and purification system. J. Chromatogr. 318:209 (1985).

15. T. Nagata and M. Saeki. Determination of ampicillin residues in fish tissues by liquid chromatography. J. Assoc. Off. Anal. Chem. 69:448 (1986).

16. K. Tyczkowska, R. D. Voyksner, and A. L. Aronson. Development of an analytical method for penicillin G in bovine milk by liquid chromatography with ultraviolet-visible detection and confirmation by mass-spectrometric detection. J. Chromatgr. 490:101 (1989).

17. R. D. Voyksner, K. L. Tyczkowska, and A. L. Aronson. Development of analytical methods for same penicillins in bovine milk by ion-paired chromatography and confirmation by mass spectrometry. J. Chromatogr. 567:389 (1991).

18. K. L. Tyczkowska, R. D. Voyksner, and A. L. Aronson. Development of an analytical method for cephapirin and its metabolite in bovine milk and serum by liquid chromatography with UV-VIS detection and confirmation by thermospray mass spectrometry. J. Vet. Pharmacol. Therap. 14:51 (1991).

19. A. I. MacIntosh. Liquid chromatographic determinations of cephapirin residues in milk. J. Assoc. Off. Anal. Chem. 73:880 (1990).

20. W. A. Moats. Determination of penicillin G in milk by high-performance liquid chromatography with automated liquid chromatographic cleanup. J. Chromatogr. 507:177 (1990).

21. W. A. Moats and R. Malisch. Determination of cloxacillin and penicillin V in milk using an automated LC cleanup. J. Assoc. Off. Anal. Chem. (In press).

22. W. A. Moats. HPLC determination of penicillin G, penicillin V, and cloxacillin in beef and pork tissues. J. Chromatogr. (In press).

23. B. Wiese and K. Martin. Determination of benzylpenicillin in milk at the pg ml^{-1} level by reversed-phase liquid chromatography in combination with digital subtraction chromatography technique. J. Pharm. Biomed. Anal. 7:95 (1989).

24. W. A. Moats. Determination of ampicillin in milk by liquid chromatography using an automated LC cleanup, in: "Residues of Veterinary Drugs in Foods", N. Haagsma, A. Ruiter, and P. B. Czedik-Eysenberg, Ed. Noordwijkerhout, The Netherlands, 1990, p 280.

25. J. O. Boison, C. D. C. Salisbury, W. Chan, and J. D. MacNeil. The determination of penicillin G residues in edible animal tissues by liquid chromatogr. J. Assoc. Off. Anal. Chem. 74:497 (1991).

26. B. Wiese and K. Martin. Basic extraction studies of benzylpenicillin and its determination by liquid chromatography with pre-column derivatization. J. Pharm. Biomed. Anal. 7:67 (1989).

27. J. P. Hou and J. W. Poole. B-lactam antibiotics: their physicochemical properties and biological activities in relation to structure. J. Pharm. Sci. 60:503 (1971).

28. T. Nakagaw, A. Shibukawa, and T. Uno. Liquid chromatography with crown ether-containing mobile phases II. Retention behavior of B-lactam antibiotics in reversed-phase high-performance liquid chromatography. J. Chromatogr. 239:695 (1982).

COMPARISON OF CHROMATOGRAPHIC PROCEDURES FOR DETERMINING

RESIDUES OF PENICILLINS

Michael Petz

University of Wuppertal
Department of Food Chemistry - FB 9
Gaußstraße 20, D-5600 Wuppertal 1

INTRODUCTION

Residues of penicillins in milk or tissues are traditionally analyzed by microbiological tests. However, these techniques can give false-positive results caused by active metabolites or antibacterial endo- or exogenic substances that may interfere in microbiological methods. Another drawback is poor precision. While bioassay procedures can distinguish resulting penicillin residues from other antibiotics by use of penicillinase, they cannot distinguish penicillins from one other. By the introduction of chromatographic procedures it is possible to separate and identify different penicillins. The main problem is that if they are to be of any value as confirmatory methods they must equal or exceed the sensitivity of the microbiological screening tests at least as long as maximum residue limits are set at these levels.

Benzylpenicillin (Penicillin G, Pen G) is the oldest antibiotic and still enjoys extensive clinical utility for treatment of mastitis and other animal diseases. Ampicillin expanded the therapeutic activity also against gram-negative organisms and the advantage of cloxacillin is the activity against penicillinase-producing bacteria (Fig. 1). Other less frequently used penicillins include phenoxymethylpenicillin (Penicillin V), methicillin, oxacillin, dicloxacillin, nafcillin and amoxicillin. Residue testing programs indicate that penicillins, tetracyclines and streptomycin rank first when positive results from microbiological tests were further investigated, for example by bioautography (1) (Fig. 2).

Chromatographic procedures have the inherent advantage of being principally capable of simultaneous multi-residue identification and quantitation. However, the majority of residue analytical methods for ß-lactams is centered on Pen G only. As far as multi-component procedures are published these are typically restricted to the seven or eight major penicillins with neutral side-chain (2-4). This is less a problem of chromatographic separation but of the extraction and cleanup procedure which in most cases exclude the amphoteric ampicillin and amoxicillin. Unconventional residue analytical cleanup techniques like the matrix solid phase dispersion approach (5), ultrafiltration (6), HPLC column switching (7) or collecting of fractions from a HPLC cleanup (8) will have the potential for being universally applicable as long as the requirements for sensitivity can be met.

Analysis of Antibiotic Drug Residues in Food Products of Animal Origin
Edited by V.K. Agarwal, Plenum Press, New York, 1992

147

Benzylpenicillin
Ampicillin
Cloxacillin

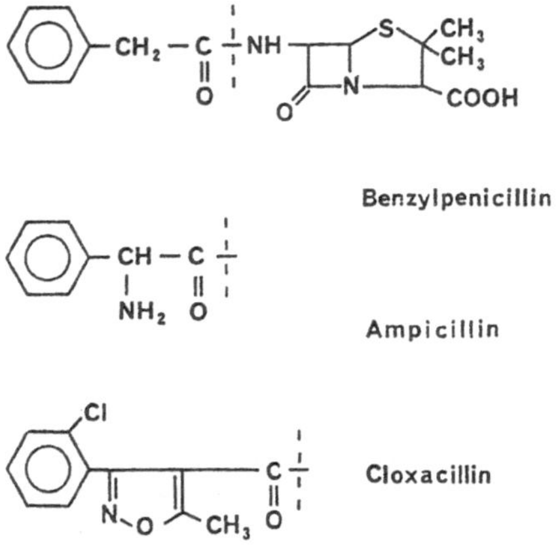

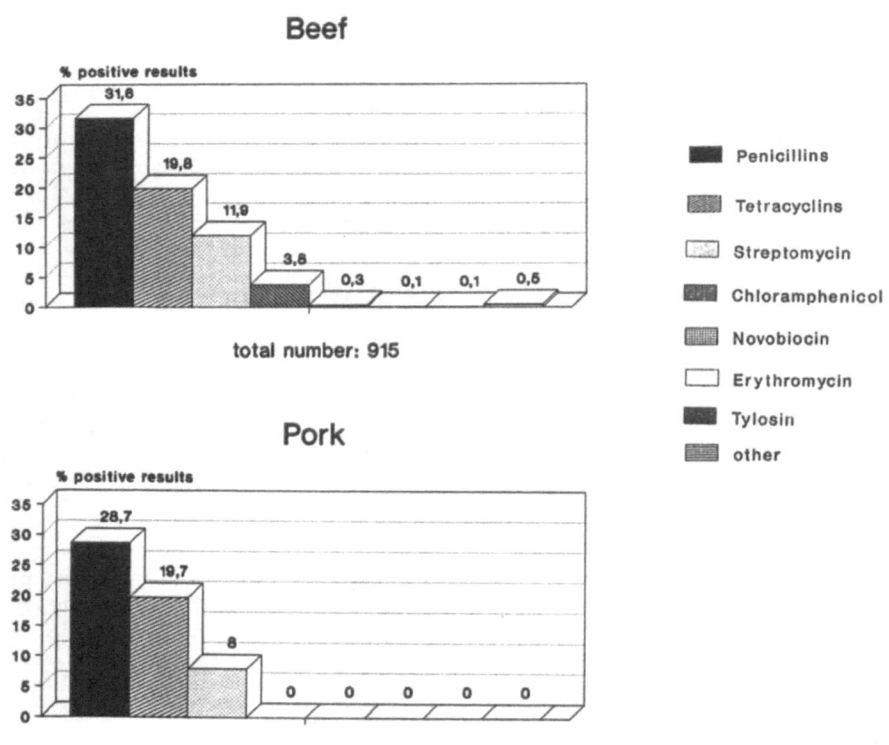

Fig. 1. Structures of extensively used penicillins

Beef
% positive results
31,6
19,8
11,9
3,8
0,3
0,1
0,1
0,5
total number: 915
Pork
% positive results
28,7
19,7
8
0
0
0
0
0
total number: 401
Penicillins
Tetracyclins
Streptomycin
Chloramphenicol
Novobiocin
Erythromycin
Tylosin
other

Fig. 2. Results of a Canadian testing program 1984-1987, *according to (1)*

With about all published methods the first step is to deproteinize the biological material either by tungstic acid or by acetonitrile prior to extraction with other organic solvents (Fig. 3). Deproteinization helps to prevent formation of emulsions during succeeding partitioning steps. Since strongly acidic deproteinizing agents cannot be used because of degradation of penicillins, tungstic acid, which gives a pH in the supernatant of about 2.2-3.9 (9) is used for effectively precipitating proteins and fats. However, Moats (10) observed erratic recoveries when milk was treated with tungstic acid. Poor recoveries apparently resulted from coprecipitation of penicillins with milk proteins since penicillins added to the filtrate were quantitatively recovered. Acetonitrile proved satisfactory for precipitating proteins and, since it is a poor solvent for fats, most of the lipid present was coprecipitated with the proteins. For recovery of penicillins from milk, a ratio of acetonitrile to milk of 2:1 was optimal. A 1:1 ratio was ineffective in precipitating proteins, and recoveries of penicillin were poor when a 3:1 ratio was used. At the higher proportion of acetonitrile, penicillins apparently coprecipitated with the protein and lipid, which formed a sticky mass at the bottom of the flask (10). Wiese and Martin (11) recommend to extract milk with acetonitrile at 30 °C and not at room temperature. At the elevated temperature the precipitated proteins and lipids gave a firmer plug after centrifugation.

Tyczkowska et al. (6) used ultrafiltration with a 10 000 Dalton cutoff filter for deproteinizing milk samples. With the ultrafiltration cleanup there is no need for transfer into an organic solvent and therefore, principally the aqueous solution could be held at the pH of optimal stability. However, Pen G, like other ß-lactam antibiotics, exhibits significant binding to proteins and other matrix components of milk, thus reducing the recovery to less than 50 % when water or aqueous buffer is used without addition of drug releasing solvents. An aqueous mixture with 40 % acetonitrile and 20 % methanol was optimal for drug release and resulted in a recovery above 80 % from milk. The evaluation of various drug releasing solvents shows the importance of protein binding on sample recoveries (Fig. 4).

Residue analysis of benzylpenicillin

EXTRACTION / DEPROTEINATION	Acetonitrile Sodium tungstate/sulfuric acid
CLEANUP	Partitioning Alumina, C-18, HPLC column-switching Matrix solid phase dispersion Ultrafiltration HPLC fractionation C-18 Anion exchange, Diol-SPE
DERIVATIZATION	Penase, HgCl2/Dansylhydrazone Penase, Imidazole/HgCl2 NaOH, Triazole/HgCl2 Methylation
CHROMATOGRAPHY / DETECTION	HPLC (UV: 200-230 nm/PDA) HPLC (FLU: 254/500 nm) HPLC (UV: 325 nm) GC (TSD)

Fig. 3. Typical steps in the chromatographic residue analysis of benzyl-penicillin. SPE = solid phase extraction, PDA = photodiodearray detector, FLU = fluorescence detector, TSD = thermionic nitrogen-selective detector

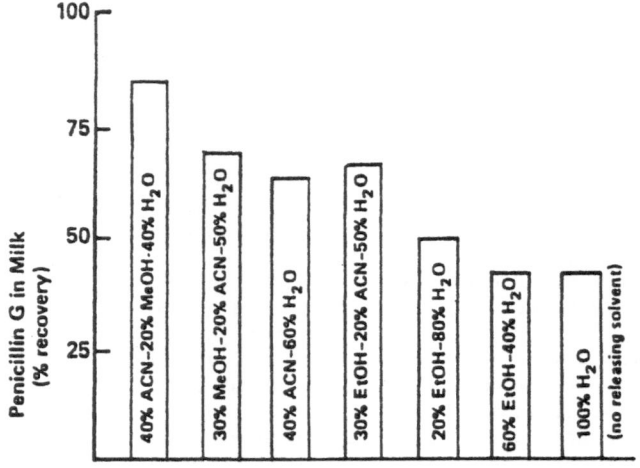

Fig. 4.   Evaluation of various solvents for their efficiency in releasing
Pen G from protein binding in milk. *From (6), with permission of
Elsevier Science Publishers.*

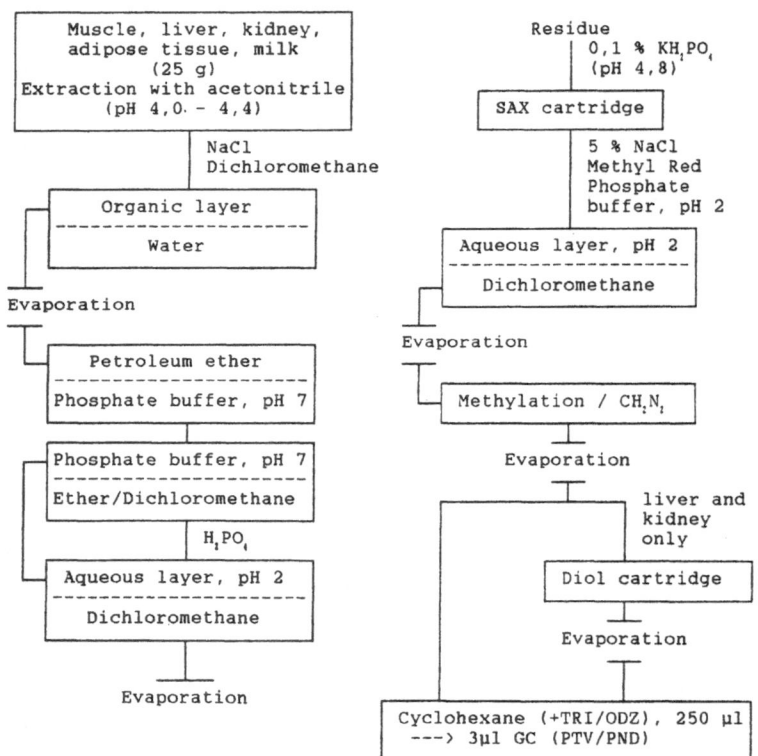

Fig. 5.   Extraction and cleanup scheme for gaschromatographic residue
analysis of penicillins with neutral side-chain. *From (4)*

The analysis of penicillin residues by chemical methods is rather delicate as they are liable to get lost during the various steps of extraction, partitioning or evaporation because of hydrolysis, alcoholysis, or irreversible adsorption. These effects increase substantially with decreasing concentrations of the penicillins in the sample. A rather fast degradation is being catalysed by both, acids and bases. Optimum stability for monobasic penicillins generally occurs at a pH between 6 and 7, while for the amphoteric penicillins it coincides with the isoelectric point (14). The instability of penicillins in acid media requires that analytical steps involving contact between acids and penicillins should be performed in a highly reproducible fashion for a minimum length of time. However, although Pen G is unstable at low pH, there was no evidence of decomposition during the HPLC analysis even when longer gradients were used with 0.01 M orthophosphoric acid, pH 1.96 (8). This is explained by the presence of the organic modifier in the mobile phase.

Because neutral penicillins are acids with pK values of about 2.6 there must be a balance between good extraction efficiency and losses of the substance caused by acidic degradation. Fig. 6 shows that a few minutes were sufficient to markedly degrade Pen G dissolved in buffer solutions of pH values between 2 and 3 (15). When sodium ions are present they allow some kind of ion pair extraction and therefore less acidic pH values are possible without a decrease in extraction efficiency.

We observed this effect with the extraction for our gas chromatographic procedure (4). When the aqueous phase from meat had a pH value of pH 4.2 (after phases separated) all investigated penicillins were about quantitatively extracted without degradation. At pH 2.6 the acid-labile penicillins Pen G, methicillin and nafcillin disappeared about completely while the acid-resistant compounds like penicillin V or the isoxazolyl penicillins survived (Fig. 7). Because the various penicillins behave differently with regard to their stability it seems critical to use one penicillin as internal standard for the analysis of another.

Methanol should be avoided as solvent because it degrades penicillins to the corresponding alkyl-α-D-penicilloic acids. However, the reaction can be slowed down by adding water or buffer to methanol, thus allowing its use in mobile phases for HPLC (16,17).

Penicillins also have the tendency to adsorb at certain interfaces. The losses that occur during evaporation steps seem to us a combination of irreversible adsorption to the used glassware and degradation that might be caused by the concentration of co-extracted components of the matrix or of concentrated traces of acids or other chemicals used for the preceding analytical steps. We found it necessary to silylate all glassware used for standards and evaporation steps. But despite this, we observed a substantial reduction with about all penicillins when standards were taken to dryness using a rotary evaporator at 30 °C. The rather inert internal standards for chromatography were not effected (Fig. 8). The adsorption of penicillins to the glassware could be a possible cause for memory effects which sometimes can be observed. It is advisable to have separate glassware for use with high concentrations (method development, standard solutions) and low concentrations. Rinsing glassware with 0.01 M hydrochloric acid is recommended as a means to avoid erratic results (8). There is obviously an advantage for those extraction and cleanup procedures that do not need to evaporate extracts to dryness. In those cases where it cannot be avoided -

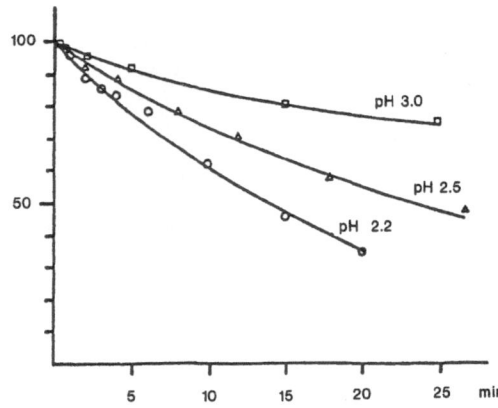

Fig. 6. Instability of Pen G at room temperature in buffer solutions of pH 2.2, pH 2.5 and pH 3.0 (100 μg/ml Pen G). *From (15), with permission of Pergamon press PLC.*

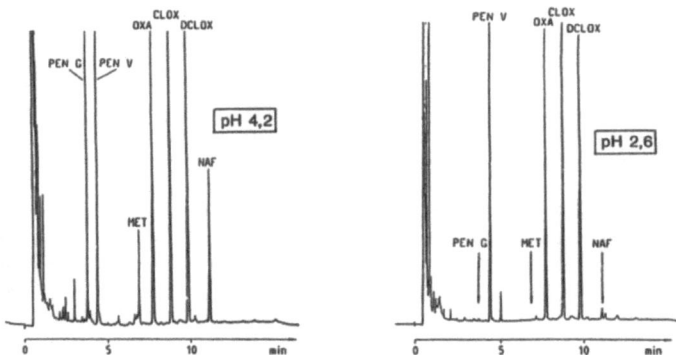

Fig. 7. Effect of extraction pH (= pH* in sodium chloride saturated aqueous phase after separation from acetonitrile/dichloromethane) on penicillin recovery from meat (500 μg/kg). *From (34)*

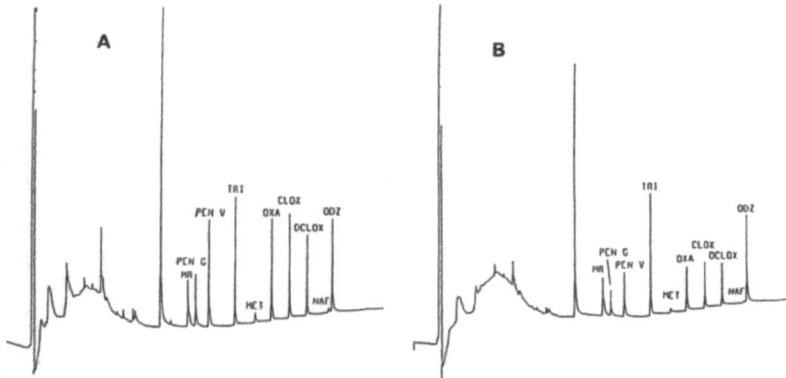

Fig. 8. Effect of evaporation on penicillin recovery. A: Rotary evaporation of extract to 4 ml (30 °C) and further reduction to 300 μl by a stream of nitrogen and hand warmth under observation. B: Rotary evaporation to dryness (30 °C); TRI/ODZ: internal standards. *From (4)*

as it is the case at 4 steps with the cleanup for the GC method (Fig. 5) we were able to recover the penicillins about quantitatively when the last ca. 300 µl are taken to dryness under observation by a stream of nitrogen and using hand warmth. The succeeding dissolution had to be aided by an ultrasonic treatment. Another suggestion to avoid adsorption of residual penicillins is to add a large excess of a related pro-drug. The use of plastic ware did not appear to diminish adsorption by our and other authors experience (14).

The analysis for penicillin residues should be performed as soon as possible after sampling or samples should be stored at a temperature as low as possible. A rather fast degradation could be observed within hours or days, respectively, when the samples were held at room or refrigerator temperatures. Even when samples were held frozen at - 18 °C there was a slow but steady decrease of Pen G (11) (Fig. 9).

DERIVATIZATION

Most methods use the native UV absorptivity of Pen G between 200 and 230 nm for detection after HPLC separation. However, a lot of endogeneous substances in milk or tissues also absorb in this region. Therefore, the required sensitivity can only be obtained by this approach after con- siderable concentration and very effective cleanup. A higher specifity and sensitivity can be obtained by reacting penicillins with derivatizing agents. Both the selectivity of the derivatization reaction and the higher detection wavelength (or the resulting fluorescence) minimise the risk of interference from endogeneous compounds. Two types of derivatization reactions have been published for residue analysis both of which use in its first step a cleavage of the ß-lactam ring (Fig. 10).

With a triazole or imidazole the ß-lactam ring is opened to form an intermediate that reacts with mercury chloride to the mercuric mercaptide of the penicillenic acid of benzyl- or another penicillin while the triazole or imidazole migrates from the molecule. The molar absorbance of the deriva- tives at 325 nm does not improve the sensitivity of penicillin detection when compared with the innate penicillin absorption at about 200 nm. However, the detection is far more selective and matrix interferences are reduced (11,13,15).

The second derivatization process in residue analysis involves in its first step the hydrolysis of the ß-lactam ring by penicillinase which is preferrable over sodium hydroxide because of penicillin degradation. Without the catalytic effect of a triazole or imidazole the penicilloic acid reacts with mercury chloride to form via the penaldic acid and penicillamine the penilloaldehyde that can be reacted with dansylhydrazine to fluorolabelled products. This method is suggested for 8 neutral penicillins (2) but is of limited value for the isoxazolyl group (oxa-, cloxa- and dicloxacillin) because these can be only partially hydrolysed by penicillinase even at elevated temperatures.

Other derivatization reactions are described for the determination of penicillins in pharmaceutical formulations, fermentation broths or biologi- cal fluids with the potential of adaptation for residue analysis. Penicil- lins with a primary amino group in the side chain have been analyzed after post-column reaction with fluorescamine (18) or by the fluorescence which developed after on-line post-column electrochemical oxidation (19). Other penicillins were derivatized by the use of 9-fluorenylmethyl-chloroformate (20), o-phthaldialdehyde (21) or 7-fluoro-4-nitrobenzo-2-oxa-1,3-di-

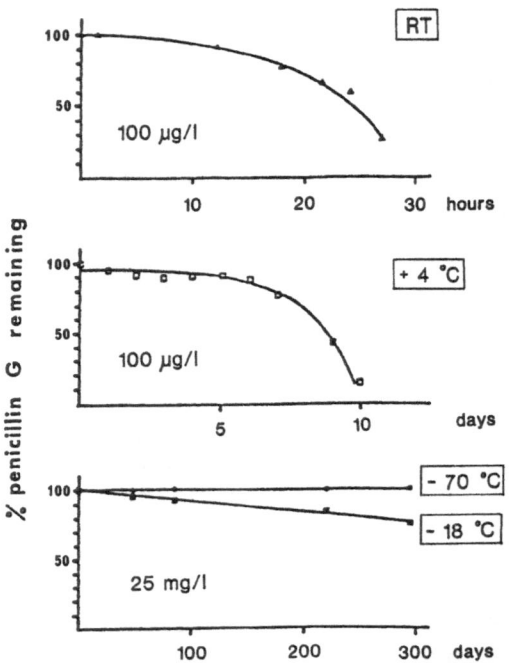

Fig. 9. Stability of Pen G in milk at room temperature (RT), + 4 °C, - 18 °C and - 70 °C. *From (11), with permission of Pergamon press PLC.*

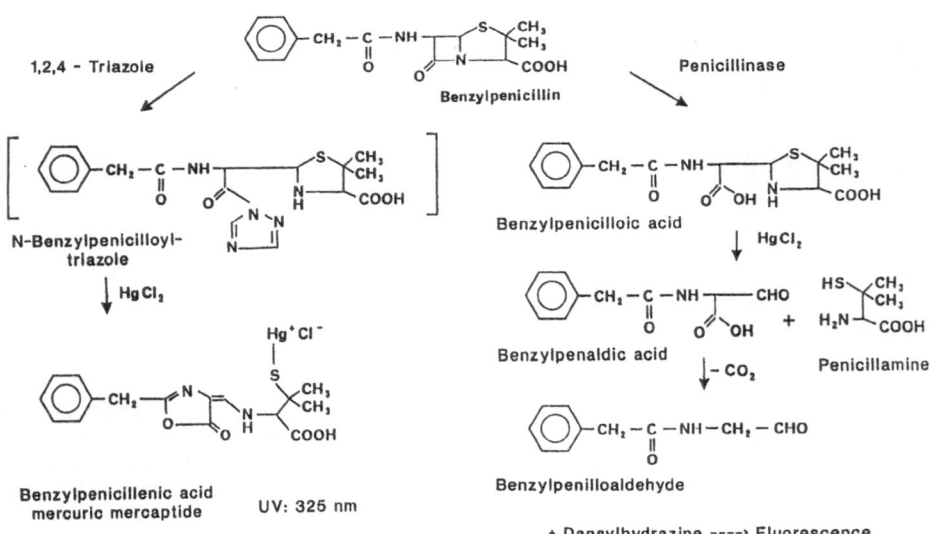

Fig. 10. Reactions used for residue analysis of Pen G

azole (NBD) (22) for fluorescence detection. The reaction with hydroxy-acetophenones (23) or with light in an on-line photo-reactor (24) enhanced the sensitivity for electrochemical detection.

The use of 4-bromomethyl-7-methoxycoumarin (BrMMC) as fluorescence label allows the detection and quantitation of penicillins in the pg-range (25). These can be extracted from aqueous solutions of ca. pH 7 into dichloromethane by the aid of 18-crown-6 ether, thus avoiding the problems of phase transfers at low pH. The ether further reacts as catalyst for the esterification which gives products with intact penicillin structure and emission maxima of about 400 nm (excitation maxima 320 nm) (Fig. 11). The ester formation was estimated with the use of ^{14}C benzylpenicillin and yielded 90 % (1 h, 55 °C) with a reproducibility (coefficient of variation) between 3.1 % and 6.7 % for the various penicillins and concentrations. This derivatization procedure is intended to be used after an immunoaffinity chromatographic cleanup which is under development. However, the selectivity and sensitivity of this type of derivatization could already be demonstrated by the exemplary analysis of a milk sample spiked at 10 μg/kg. The cleanup for this sample was limited to the partitioning steps of the GC method (Fig. 5). As can be seen, there is no interference from the matrix even with this minimized cleanup and the sensitivity is sufficient to quantitate residues at the 4 μg/kg level (Fig. 12). A further advantage is, that the derivatives have a better retention on C-18 columns and the separation of the penicillins is possible without a gradient, without ion-pairing and at the pH of optimal stability.

CHROMATOGRAPHY

A number of papers report work carried out to compare the chromato-graphic behaviour of penicillins and their degradation products with respect to column type and temperature, to pH, buffer concentration, ion-pairing reagents, organic modifier type and content in the mobile phase (for example 14, 26-30).

There is no clear preference for either methanol or acetonitrile as organic modifier on either C-18, C-8 or phenyl columns. With ion-pair reversed-phase chromatographic procedures typically the tetraethyl- or butylammonium cation has been used as the counter ion to increase retention time. But conversion of penicillins from the ionized form to the acidic form at pH 1.96 produced a much larger retention on a RP column than ion-pairing and thus good separation from interferences (8). The addition of amines to the mobile phase has been shown to improve the selectivity in ampicillin analysis and can also be generally used to suppress silanol effects. Similarily, alkylsulfonic acids have been used to improve the separation not only of penicillins with an amine function in their side-chain (14) but also with neutral penicillins (6). Moats found that polymeric columns were more stable than silica-based RP columns in the pH range of 7-8. He observed that under his conditions PLRP-S columns approached column efficiencies of bonded silica packings of comparable size, but contrary to the manu-facturer's claims these packings were not stable to pH 13 and developed excessive back pressures above pH 8 indicating the packing swelled (8).

Reversed phase HPLC methods mainly use isocratic elution. Since penicillins have a wide range of polarity from the very polar amoxicillin to the relatively unpolar dicloxacillin only a few can be recovered in a reasonable length of time at any given solvent strength. However, all can be eluted and separated nicely with an acetonitrile gradient (26).

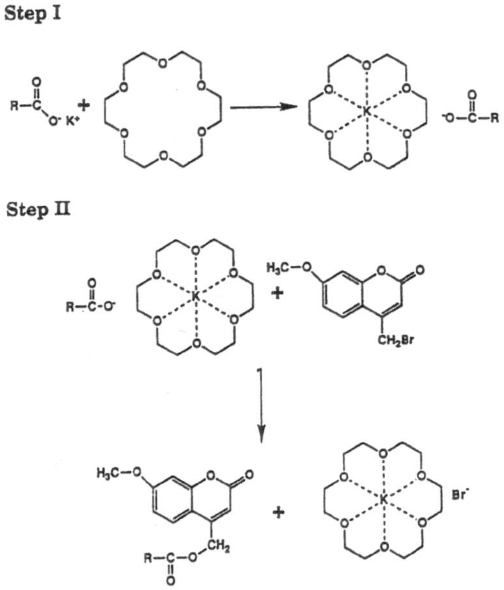

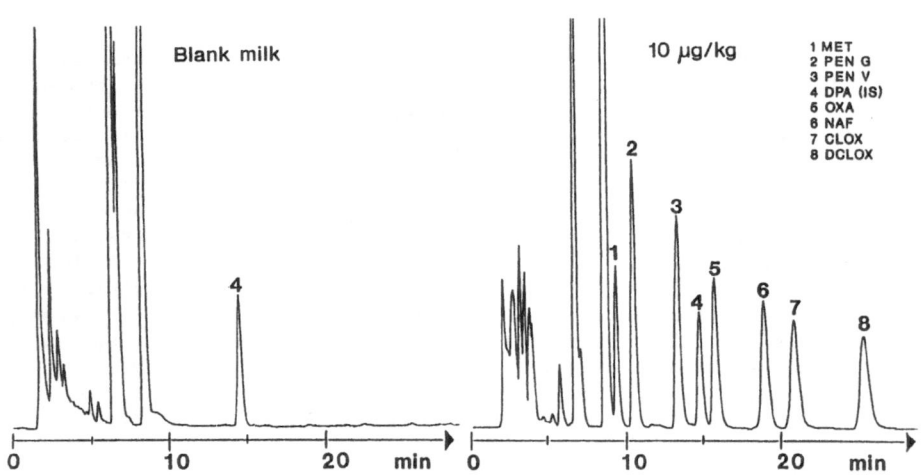

Step I

Step II

Fig. 11: Fluorescence labelling of intact penicillins. Step I: activation of the carboxyl group (R - penicillin) by crown ether. Step II: ester formation by the reaction of the activated carboxyl group with BrMMC. *From (25)*

Blank milk

10 µg/kg

1 MET
2 PEN G
3 PEN V
4 DPA (IS)
5 OXA
6 NAF
7 CLOX
8 DCLOX

Fig. 12: Liquid chromatogram of fluorolabelled penicillins after extraction from milk. *From (25)*

It is advantageous to carry out the separation at elevated temperatures. An increase in temperature for example from 20 °C to 50 °C will
shorten the analysis time and reduce the back-pressure of the column if the
flow-rate is kept constant. Sometimes even the plate number was slightly
increased (16). Wiese and Martin demonstrated that it is necessary to keep
the connection between column and detector at the same elevated temperature
to avoid a drop in column efficiency (15).

The second powerful technique besides HPLC for residue analysis of
penicillins is gas chromatography (GC). Hamann et al. (31) demonstrated
already in 1975 that residues of isoxazolyl penicillins in milk could be
analyzed by GC with packed columns and electron capture detection after
methylation of the antibiotics, although peak shape and separation from
interfering compounds were not fully satisfying. Since that time, tech-
nologies of GC column production, injection devices and detectors improved
markedly. With fused silica capillary columns coated with very thin films
(0.1 μm) of 100 % methyl silicone, separation of the methylated penicillins
was excellent (Fig. 6). Cold splitless, or alternatively on-column injection
in connection with the use of thermionic nitrogen-selective detection (TSD
or PND) were necessary for the development of a highly sensitive residue
analytical procedure (3,4).

RESIDUE METHOD CHARACTERISTICS AND PERFORMANCES

Moats has been the first who published physico-chemical methods for
residue analysis of Pen G with validation data for the quantitative
determination of this antibiotic in milk and tissues in '83 (32) and '84
(35). He used the innate UV absorption at 210 nm after a partitioning
cleanup and HPLC separation. During the following years other authors
suggested alternative HPLC methods with cleanup and derivatization procedures mentioned above (Fig. 13). However, it was not before 1989 that a

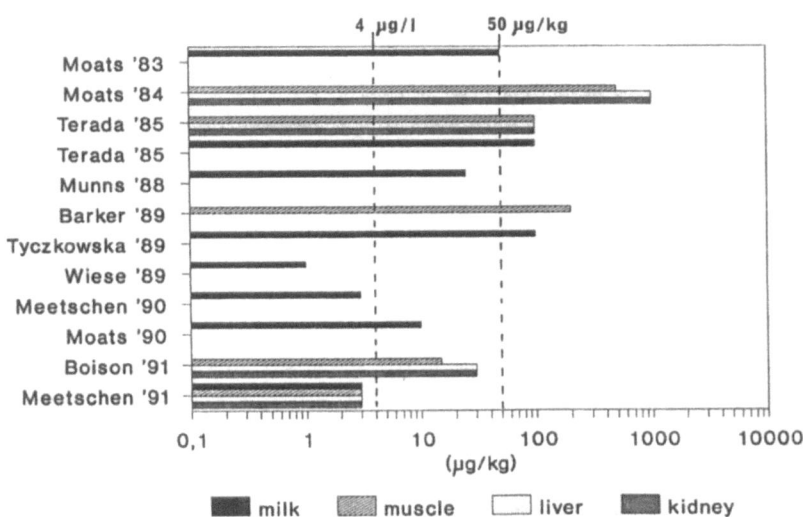

Fig. 13: Performance characteristics of residue analytical methods for
Pen G in milk and tissues with regard to lowest quantitated
concentrations

method has been published which demonstrated its applicability for quantitative analysis at or below the sensitivity of the microbiological tests of about 3 or 4 parts per billion in milk or tissues. Now this value of 4 µg/kg has become internationally recommended as maximum residue limit for milk by the FAO/WHO Joint Expert Committee on Food Additives (JECFA) as the advisary body to the Codex Alimentarius Committee on "Residues of Veterinary Drugs in Food" and also within the European Community. A decision limit of 3 µg/kg for acceptance or condemnation of a carcass is still valid in Germany, but internationally a value of 50 µg/kg seems to be established as maximum residue limit - so by JECFA and within the EC. Till now, only two methods for milk and two methods for tissues demonstrated their quantitative performance data at or below these levels. The methods of Wiese and Martin (11) and the method of Boison et al. (13) use the formation of the mercuric mercaptide with HPLC/UV analysis while Meetschen and Petz reported a gas chromatographic procedure with thermionic nitrogen-selective detection of the methylated penicillins (3,4). All available residue analytical methods for penicillins are restricted to or include Pen G with two exceptions, the old GC method for isoxazolyl penicillins in milk (31) and for ampicillin in fish (33).

Fig. 14 presents the chromatographic results of the Wiese/Martin method. The identification of the presence of penicillin residues and the quantitation became possible only due to electronic subtraction of the matrix interferences. To achieve this, the final extract of a sample (225 µl) was divided in two equal parts. One part was first treated with penicillinase while the other was first treated with derivatization reagent. In the sample with the enzyme treatment the degraded penicillin did not react with the derivatization reagent and in the other sample with the derivatized penicillin this was not attacked by penicillinase. Boison et al. were successful quantitating tissues below 50 µg/kg without what Wiese and Martin called "Digital subtraction technique", but they did not present chromatograms at this concentration level.

Fig. 15 presents the gas chromatographic result of the analysis of a kidney sample spiked with 10 µg/kg of each penicillin. The corresponding blank shows the signals of the two internal standards. All penicillins are more than base-line separated and even at this concentration level there is hardly any interference of matrix components. We validated this gas chromatographic procedure after fortifying milk, muscle and other tissues at 10 and 3 µg/kg. Depending on the matrix and the particular penicillin the results for recovery were between 50 and 80 % without major difference between 10 and 3 µg/kg validation. With the exception of nafcillin in liver the precision of the method expressed as coefficient of variation was better than ± 20 %, in most cases even better than 10 % (Fig. 16)

With kidney or liver it was necessary to add a further solid phase extraction step with diol cartridges after the methylation to separate the methylated penicillins from interfering compounds from the matrix that survived the preceding cleanup. It was interesting that these matrix compounds did not directly interfere with the chromatography when an extract without diol cartridge cleanup was injected on a new or cleaned column. But with the second or further injections the interference became obvious with dramatically decreased responses for the penicillins while the rather inert internal standards remained unchanged. However, it is a severe drawback of the gas chromatographic procedure that it needs four evaporation steps, ion exchange chromatography, diol solid phase extraction, methylation and a specialized GC equipment.

OUTLOOK

The main purpose of physico-chemical methods is to unambiguously

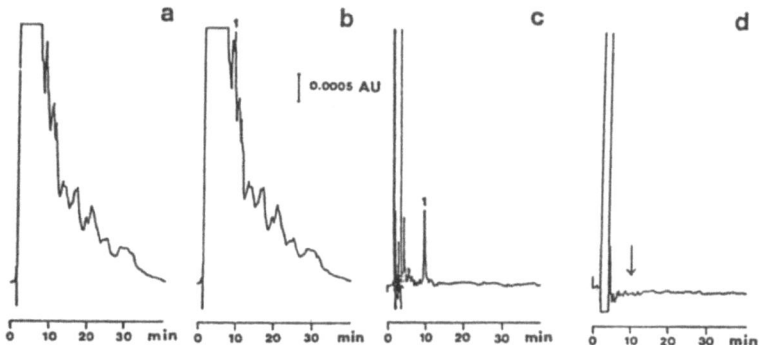

Fig. 14: Chromatograms from human milk with 2 µg/l Pen G (a,b,c): (a) Penicillinase treated sample (zero sample); (b) actual sample; (c) interfering peaks removed by the use of the digital subtraction technique, reintegrated and replotted (a minus b), 1 = Pen G; (d) blank milk performed according to (c). *From (11), with permission of Pergamon press PLC.*

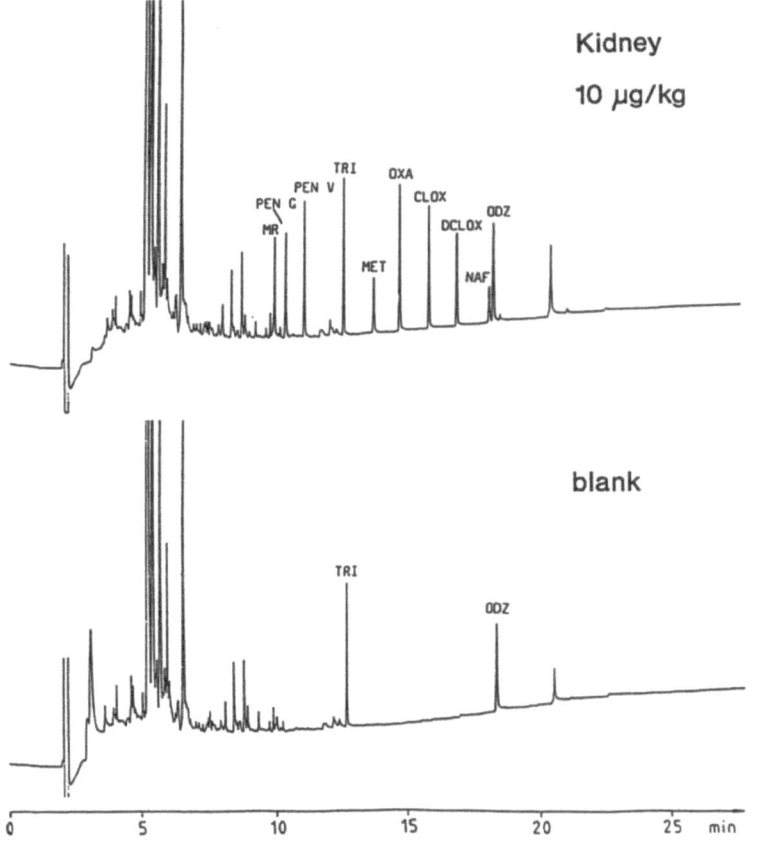

Fig. 15: GC analysis of a kidney sample spiked with 10 µg/kg. *From (4)*

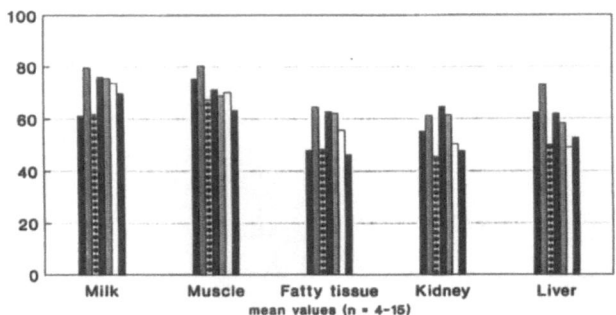

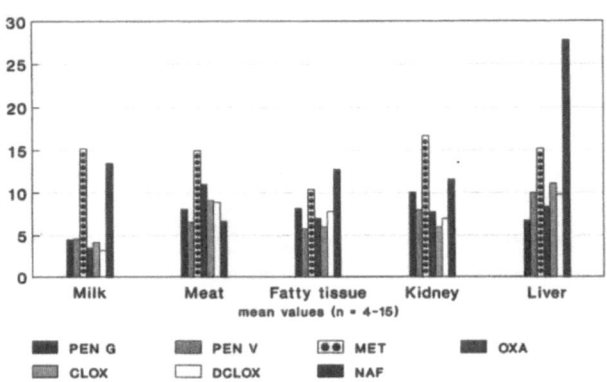

Fig. 16. Method performance data of the GC procedure. *From (4)*

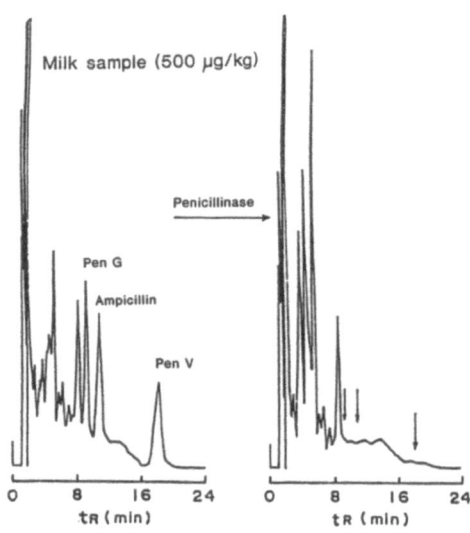

Fig. 17. Confirmation of penicillin residues by the use of penicillinase.
From (12), with permission of Elsevier Science Publishers.

confirm positive results obtained by fast, but less specific screening tests and to act as reference methods. Currently, most procedures for residue analysis of penicillins are either of insufficient sensitivity or practicability. It can be expected that future chromatographic methods will use building stones which partly are already available but not yet put together; i.e. a combination of a fast and effective cleanup step (MSPD, HPLC fractionation or immunoaffinity chromatography) with a highly selective and sensitive detection (fluorescence after derivatization and HPLC or GC with thermionic or mass selective detection). According to EC regulations on reference methods, the identification of a suspect compound should preferably be based on molecular spectroscopic data. One way to confirm chromatographic peaks as penicillins is to use penicillinase (Fig. 17). Procedures that generate molecular spectroscopic data for residues of penicillins at 4 µg/kg (milk) or 50 µg/kg (tissues) by LC-MS or GC-MS are not yet available. The only published confirmation by LC-MS is limited to concentrations above 100 µg/kg Pen G in milk (6). But more sensitive mass spectrometric methods for residue analysis of penicillins can be expected in the near future. Work is carried out on this at least in two research groups, one on LC-MS (13) and one on GC-MS (3).

REFERENCES

1. C. D. Salisbury, C. E. Rigby, and W. Chan, Determination of antibiotic residues in Canadian slaughter animals by thin-layer chromatography-bioautography, J. Agric. Food Chem. 37:105 (1989).

2. R. K. Munns, W. Shimoda, J. E. Roybal, and C. Vieira, Multiresidue method for determination of eight neutral ß-lactam penicillins in milk by fluorescence-liquid chromatography, J. Assoc. Off. Anal. Chem. 68:968 (1985).

3. U. Meetschen and M. Petz, Capillary gas chromatographic method for determination of benzylpenicillin and other beta lactam antibiotics in milk, J. Assoc. Off. Anal. Chem. 73:373 (1990).

4. U. Meetschen and M. Petz, Gas chromatographic method for the determination of residues of seven penicillins in foodstuffs of animal origin, Z. Lebensm. Unters. Forsch. 194 (4) (1991) in print.

5. S. A. Barker, A. R. Long and C. R. Short, Isolation of drug residues from tissues by solid phase dispersion, J. Chromatogr. 475:353 (1989).

6. K. Tyczkowska, R. D. Voyksner, and A. L. Aronson, Development of an analytical method for penicillin G in bovine milk by liquid chromatography with ultraviolet-visible detection and confirmation by mass-spectrometric detection, J. Chromatogr. 490:101 (1989).

7. H. Terada, M. Asanoma, and Y. Sakabe, III. High-performance liquid chromatographic determination of penicillin G in animal tissues using an on-line pre-column concentration and purification system, J. Chromatogr. 318:299 (1985).

8. W. A. Moats, Determination of penicillin G in milk by high-performance liquid chromatography with automated liquid chromatographic cleanup, J. Chromatogr. 507:177 (1990).

9. H. Lingeman, R. D. McDowall, and U. A. Th. Brinkman, Guidelines for bioanalysis using column liquid chromatography, Trends in Analytical Chemistry 10:48 (1991).

10. W. A. Moats, Detection and semiquantitative estimation of penicillin G and cloxacillin in milk by thin-layer chromatography, *J. Agric. Food Chem*. 31:1348 (1983).

11. B. Wiese and K. Martin, Determination benzylpenicillin in milk at the pg ml^{-1} level by reversed-phase liquid chromatography in combination with digital subtraction chromatography technique, *J. Pharm. Biomed. Anal*. 7:95 (1989).

12. H. Terada and Y. Sakabe, IV. Simultaneous determination of penicillin G, penicillin V and ampicillin in milk by high-performance liquid chromatography, *J. Chromatogr*. 348:379 (1985).

13. J. O. Boison, C. D. C. Salisbury, W. Chan, and J. D. MacNeil, Determination of penicillin G residues in edible animal tissues by liquid chromatography, *J. Assoc. Off. Anal. Chem*. 74:497 (1991).

14. J. O. Miners, The analysis of penicillins in biological fluids and pharmaceutical preparations by high-performance liquid chromatography: A review, *J. Liquid Chromatogr*. 8:2827 (1985).

15. B. Wiese and K. Martin, Basic extraction studies of benzylpenicillin and its determination by liquid chromatography with pre-column derivatisation, *J. Pharm. Biomed. Anal*. 7:67 (1989).

16. F. Nachtmann and K. Gstrein, Simultaneous determination of the cationic and anionic parts in respiratory penicillins by high-performance liquid chromatography, *J. Chromatogr*. 236:461 (1982).

17. U. R. Tjaden, H. Lingeman, R. A. M. van der Hoeven, J. A. C. Bierman, H. J. E. M. Reeuwijk, and J. van der Greef, High-performance liquid chromatographic analysis of isoxazolylpenicillins in plasma and urine samples, *Chromatographia* 24:597 (1987).

18. J. Carlqvist and D. Westerlund, Automated determination of amoxycillin in biological fluids by column switching in ion-pair reversed-phase liquid chromatographic systems with post-column derivatization, *J. Chromatogr*. 344:285 (1985).

19. H. Mascher and C. Kikuta, Determination of amoxicillin in plasma by high-performance liquid chromatography with fluorescence detection after on-line oxidation, *J. Chromatogr*. 506:417 (1990).

20. A. J. Shah and M. W. Adlard, Determination of ß-lactams and their biosynthetic intermediates in fermentation media by pre-column derivatisation followed by fluorescence detection, *J. Chromatogr*. 424:324 (1988).

21. M. E. Rogers, M. W. Adlard, G. Saunders, and G. Holt, Derivatization techniques for high-performance liquid chromatographic analysis of ß-lactams, *J. Chromatogr*. 297:385 (1984).

22. K. Iwaki, N. Okumura, M. Yamazaki, N. Nimura, and T. Kinoshita, Precolumn derivatization technique for high-performance liquid chromatographic determination of penicillins with fluorescence detection, *J. Chromatogr*. 504:359 (1990).

23. R. K. Munns, J. E. Roybal, W. Shimoda, and J. A. Hurlbut, 1-(4-Hydroxyphenyl)-, 1-(2,4-dihydroxyphenyl)- and 1-(2,5-dihydroxyphenyl)-2-bromoethanones: New labels for determination of carboxylic acids by high-performance liquid chromatography with electrochemical and ultraviolet detection, *J. Chromatogr*. 442:209 (1988).

24. C. M. Selavka, I. S. Krull, and K. Bratin, Analysis for penicillins and cefoperazone by HPLC-photolysis-electrochemical detection (HPLC-hv-EC), _J. Pharm. Biomed. Anal_. 4:83 (1986).

25. K. Berger and M. Petz, Fluorescence HPLC determination of penicillins using 4-bromomethyl-7-methoxycoumarin as pre-column labelling agent, _Dtsch. Lebensm. Rundsch_. 87:137 (1991).

26. W. A. Moats, Effects of the silica support of bonded reversed-phase columns on chromatography of some antibiotic compounds, _J. Chromatogr_. 366:69 (1986).

27. W. A. Moats and L. Leskinen, Comparison of bonded, polymeric and silica columns for chromatography of some penicillins, _J. Chromatogr_. 386:79 (1987).

28. A. M. Lipczynski, Reversed-phase high-performance liquid chromatographic retention behaviour of benzylpenicillin and its acid-base degradation products, _Analyst_ 112:411 (1987).

29. J. Martin, R. Mendez, and A. Negro, Effect of temperature on HPLC separations of penicillins, _J. Liquid Chromatogr_. 11:1707 (1988).

30. P. C. Van Krimpen, W.P. van Bennekom, and A. Bult, Penicillins and cephalosporins - Physicochemical properties and analysis in pharmaceutical and biological matrices, _Pharm. Weekbl. Sci. Ed_. 9:1 (1987).

31. J. Hamann, A. Tolle, A. Blüthgen, and W. Heeschen, Studies on detection, isolation and identification of antibiotic residues in milk. 1. Isoxazolyl penicillins, _Milchwissenschaft_ 30:1 (1975).

32. W. A. Moats, Determination of penicillin G, penicillin V, and cloxacillin in milk by reversed-phase high-performance liquid chromatography, _J. Agric. Food Chem_. 31:880 (1983).

33. T. Nagata and M. Saeki, Determination of ampicillin residues in fish tissues by liquid chromatography, _J. Assoc. Off. Anal. Chem_. 69:448 (1986).

34. U. Meetschen, Gas chromatographic method for residues of seven penicillins in foodstuffs of animal origin. Thesis, University of Münster/Germany, 1990.

35. W.A. Moats, Determination of pencillin G and cloxacillin residues in beef and pork tissue by high-performance liquid chromatography, _J. Chromatogr_. 317:311 (1984).

A HIGH PERFORMANCE LIQUID CHROMATOGRAPHIC METHOD FOR THE
DETERMINATION OF NINE SULFONAMIDES IN MILK

Vipin K. Agarwal

The Connecticut Agricultural Experiment
Station, Box 1106, New Haven, CT 06504

ABSTRACT

A High Performance Liquid Chromatographic (HPLC) method for the determination of sulfadiazine, sulfathiazole, sulfapyridine, sulfamerazine, sulfamethiazole, sulfamethazine, sulfachloropyridazine, sulfadimethoxine, and sulfaquinoxaline in milk is described. Milk (10mL) is extracted three times (50, 40, and 40mL) with chloroform/acetone, the extract evaporated to dryness, and redissolved in 5mL potassium phosphate buffer (1.0 molar, pH 4.4). Hexane (5mL) is added to extract the lipid materials. The aqueous buffer layer containing sulfonamides is passed through a cyclobond-I solid phase extraction (SPE) cartridge and the cartridge is washed with 5mL potassium phosphate buffer. The sulfonamides, which are retained on the cartridge, are then eluted with 4mL aqueous acetonitrile (20% water). The eluent is evaporated to remove acetonitrile and the final volume of the aqueous eluent is made to 4mL with ammonium acetate buffer. The eluent is analyzed by HPLC using a reverse phase column with UV detection at 265nm. The recoveries of individual sulfonamides ranged from 64.0% to 85.9% in samples fortified at 10 and 20ppb levels.

INTRODUCTION

The use of sulfonamides as veterinary drugs for the treatment of a variety of bacterial infections is very common. In food producing animals, sulfonamides are used not only for therapeutic purposes but also for prophylactic purposes.

There has been more concern lately about the residues of sulfamethazine in milk since a study by the Food and Drug Administration's National Center for Toxicological Research (NCTR) indicated that it may be a carcinogen (1). Residues of sulfonamides in foods can be a health hazard to consumers (2). Firstly, the carcinogenicity of some sulfonamides such as sulfamethazine may be a serious concern (2). Secondly, continuous exposure of certain microorganisms to these drugs may result in the development of drug resistant strains (2). In the past, residues of sulfa drugs have been found in milk offered for sale to the consumers. A nationwide survey by the Food and Drug Administration in 1988 reported that 45% of the milk samples contained detectable amounts of sulfamethazine (3). In another survey, which included 30 samples from 10 cities across Canada, two contained sulfamethazine residues at 11.40 and 5.24ppb levels (4).

The U. S. Food and Drug Administration has, therefore, set tolerance limits of sulfonamides in milk. Presently, sulfadimethoxine is the only sulfa drug allowed for use in lactating animals and the residues of total sulfonamides, including sulfadimethoxine, should not exceed 10ppb in milk (5).

Sulfonamide residues in milk have been determined by various techniques which include immunoassay (6), microbiological (7), thin layer chromatography (TLC) (8-10), gas chromatography (11), and high performance liquid chromatography (12-20).

High performance liquid chromatography has become the most widely used technique for the analysis of sulfa drug residue in milk and a number of methods have been described (12-20). In general all HPLC methods which can detect sulfonamides to 10ppb level require extensive cleanup steps before HPLC analysis.

An HPLC method developed by Smedley and Weber (16) has successfully been applied for the detection of ten sulfonamides in milk to 10ppb level and is presently being used by the FDA for testing milk. This method involves a chloroform/acetone extraction followed by partitioning of the extract between hexane and potassium phosphate buffer before HPLC analysis. A number of extraneous peaks were, however, present in the chromatogram which could make quantitation difficult. Also, two separate chromatographic conditions are used for the determination of ten sulfonamides.

In this report an HPLC method for the quantitative determination of nine sulfonamides at a low level of 10ppb is described.

EXPERIMENTAL

Reagents

(a) Sulfonamides: Sulfadiazine (SDZ), Sulfathiazole (STZ), Sulfapyridine (SPD), Sulfamerazine (SMR), Sulfamethiazole (SMTZ), Sulfamethazine (SMZ), Sulfachloropyridazine (SCP), Sulfadimethoxine (SDM), Sulfaquinoxaline (SQX) (Sigma Chemical Co., St. Louis, MO).

(b) Potassium phosphate buffers (mono and dibasic), ammonium acetate, acetic acid, chloroform, acetone and HPLC grade methanol (Fisher Chemical Co., Fairlawn, NJ).

(c) Potassium phosphate buffer: Dissolve 13.60gm monobasic potassium phosphate buffer in 100mL distilled water.

(d) Ammonium acetate buffers: (1) To prepare a 25 millimolar ammonium acetate buffer, dissolve 1.95gm ammonium acetate in 900mL distilled water, adjust pH to 4.7 with acetic acid and make final volume to 1000mL. (2) Dissolve 1.95gm ammonium acetate in 900mL distilled water, adjust pH to 8.0 with ammonium hydroxide and make final volume to 1000ml.

(e) Mobile phase for HPLC:
 Solvent A: Ammonium acetate buffer a(pH 4.7):methanol (850:150).

Solvent B: Ammonium acetate buffer b(pH 8.0):methanol (700:300).
Use a gradient starting at 0% B to 100% B as follows.

```
Time 0 minutes -------------- 0% B
Time 5    "     -------------- 0% B
Time 15   "     -------------- 100% B
Time 30   "     -------------- 100% B
Time 40   "     -------------- 0% B
```

(f) Sulfonamide standard solutions: Dissolve 100mg of each of the nine sulfonamides in 100mL methanol. Dilute 1mL of this solution to 100mL with distilled water in a volumetric flask (Solution B).

(g) Working standard: Dilute 10mL of solution B to 100mL with distilled water to prepare a solution with a concentration of 1000ng/mL of each sulfonamide.

(h) Extraction solvent: Chloroform:acetone (2:1, v/v).

Apparatus

(a) Solid Phase Extraction Cartridges.- Cyclobond-I, SPE cartridges, 3mL size (ASTEC, Whippany, NJ).

(b) Liquid Chromatograph.- A CM 4000 Multiple solvent delivery system equipped with a 7125 Rheodyne injector and a SpectroMonitor 3100 variable wavelength UV detector was used (LDC/Milton Roy, Riviera Beach, FL). A HP3390 integrator was used for quantification by peak heights.

(c) LC Column.- LC 18-DB, 250 X 2.1 mm, 5um particle size. (Supelco Inc., Bellefonte, PA).

Fortification of Milk

Fortify milk with sulfonamides at 10 and 20ppb levels by diluting 100 and 200 uL of working standard to 10 mL with milk.

Extraction and Cleanup Procedure

Pipet 10mL milk into a 125mL separatory funnel and add 50mL of extraction solvent. Shake the mixture vigorously for one minute and allow to stand for 2-3 minutes. Repeat shaking the funnel three to four times and finally allow the two phases to separate. Filter the chloroform/acetone extract through a Whatman No.2 filter paper into a 125mL pear shaped flask. Reextract milk twice with 40mL extraction solvent each time and filter into the same pear shaped flask. Evaporate chloroform/acetone on a rotatory evaporator while maintaining the temperature below 40 C. Add 5mL potassium phosphate buffer (c) into the flask and shake vigorously on a vortex mixer for about one minute. Add 5mL hexane to the flask, shake well on vortex mixer and let stand for 4-5 minutes to allow separation of the two phases.
Condition a Cyclobond-I SPE cartridge by washing with 5mL distilled water followed by 5mL potassium phosphate buffer (c). Pass aqueous layer in the flask through the SPE cartridge. Add an additional 5mL of potassium phosphate buffer into the flask, shake on vortex mixer, and allow to stand for 4-5 minutes. Again pass aqueous layer through the same SPE cartridge. Wash the SPE cartridge with an additional 5mL buffer and then with 5mL of a mixture of buffer/distilled water (1/4). Elute sulfonamides with 4mL aqueous acetonitrile (20% water) into a pear shaped flask. Evaporate acetonitrile from the eluent using a rotatory

evaporator and transfer aqueous layer into a small test tube. Rinse
flask twice with 0.5mL ammonium acetate buffer and transfer the washings
to the same test tube. Adjust the final volume to 4mL and inject 100ul
into the HPLC.

RESULTS AND DISCUSSION

The calibration plots for the standard sulfonamides
were obtained by plotting the peak heights versus the concentrations of
sulfonamides ranging from 1 to 50ng. The plots were linear with
correlation coefficients over 0.987.

Figure 1 shows a representative chromatogram of nine sulfonamides.
Unfortified milk (10mL) was extracted with chloroform:acetone, cleaned
up on a cyclobond-I SPE cartridge as detailed in the procedure, and
examined by HPLC. The HPLC chromatogram shows no peaks eluting at the
retention times of any of sulfonamides (Figure 2). Peaks A, B, C, D, and
E are from the milk matrix and do not interfere with analysis. The milks
fortified with sulfonamides at 10 and 20ppb levels were also extracted,
cleaned-up and examined by HPLC. Figure 3 shows a representative
chromatogram of a milk sample fortified at 20 ppb level with well
resolved peaks for nine sulfonamides.

The recoveries of sulfonamides from the fortified milk samples were
calculated from peak heights (Table 1). The recoveries ranged from 64.0
% to 85.9 % in samples fortified at 10 and 20ppb levels.

Table 1. Recovery of nine sulfonamides in milk
fortified at 10 and 20 ppb.

Sulfonamides	RT min.	Fortification, ppb	
		10 Aver.± SD* %	20 Aver. ± SD %
Sulfadiazine	4.6	74.6 ± 6.3	75.8 ± 5.1
Sulfathiazole	6.4	69.3 ± 2.2	67.7 ± 3.5
Sulfapyridine	7.0	85.9 ± 8.7	81.2 ± 4.2
Sulfamerazine	8.0	71.0 ± 7.8	78.3 ± 6.0
Sulfamethiazole	11.6	81.7 ± 7.2	85.0 ± 3.8
Sulfamethazine	14.2	82.0 ± 6.7	79.8 ± 4.8
Sulfachloropyridine	17.3	64.0 ± 4.4	64.3 ± 7.0
Sulfadimethoxine	31.5	84.0 ± 6.0	81.7 ± 6.1
Sulfaquinoxaline	32.1	78.8 + 6.5	74.2 ± 3.0

* Average of three analysis for each fortification level.
RT= Retention time.

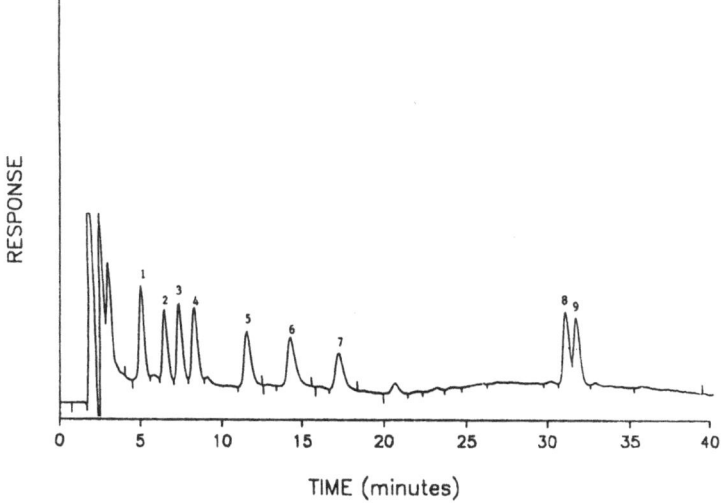

Figure 1.   Chromatogram of Standard Sulfonamides.
Sulfadiazine (1), Sulfathiazole (2),
Sulfapyridine (3), Sulfamerazine (4),
Sulfamethiazole (5), Sulfamethazine (6),
Sulfachloropyridine (7), Sulfadimethoxine(8),
Sulfaquinoxaline (9).

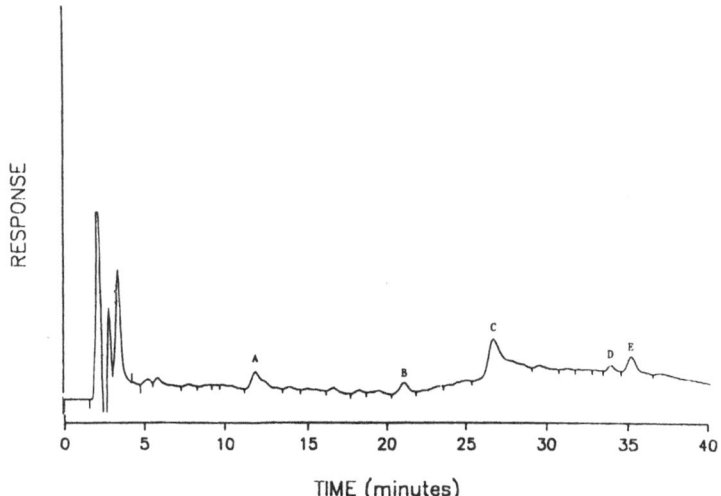

Figure 2.   Chromatogram of Blank Milk Sample. Peaks A, B, C,
D, and E are From Milk Matrix.

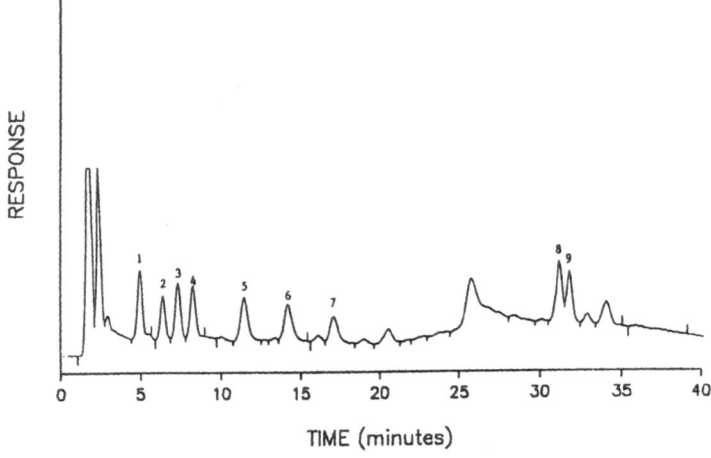

Figure 3. Chromatogram of Milk Sample Spiked with Nine
Sulfonamides Each at 20 ppb Level.

In the method presently being used by the FDA for the analysis of
sulfonamides in milk, the chloroform:acetone extract is evaporated to
dryness and the concentrated extract is partitioned between 5mL hexane
and 1mL potassium diphosphate buffer. Sulfonamides are dissolved in
buffer and are directly analyzed by HPLC. The HPLC chromatogram obtained
by this method showed numerous interferring peaks. Therefore, it was
necessary to clean the extract further to remove these interferring
materials and this was accomplished by a solid phase extraction using
cyclobond SPE cartridges.

Cyclobond-I SPE cartridges were selected to clean-up the extract
since sulfonamides in aqueous medium can easily be retained on these
cartridges in an aqueous medium. Cyclobond-I SPE cartridges contain B
Cyclodextrin bonded to 40 micron silica, which can form an inclusion
complex with sulfonamides in aqueous solution. Since sulfonamides are
ionic in nature, their retention on the cyclobond-I cartridges is pH
dependent (21). A detailed study to investigate the optimum conditions
for the formation of the inclusion complex between sulfonamide and B-
cyclodextrin has been carried out in our laboratory (22). With the
exception of sulfanilamide, all nine sulfonamides were retained within
the pH range 4.0 to 5.5. The effect of pH was more pronounced on
sulfamethazine and sulfamethiazole. Maximum retention of sulfamethazine
(87%) was obtained within the pH range 5.0 to 5.5. Sulfamethiazole
however, had a maximum retention (85%) at pH 4.0. Increasing the pH
resulted in drop of retention to almost no retention at pH 5.5.
Therefore, a potassium phosphate buffer with pH 4.4 was chosen as an
appropriate buffer. Increasing the molar strength of the buffer appeared
to give some positive effect on the retention, and one molar potassium
phosphate buffer was found to be optimum (22). Washing of SPE cartridges
with distilled water after loading the sulfonamides resulted in lower
recoveries as has been shown in the past (22). This was due to the

change in pH which affected the retention of sulfonamides on the cartridge. It was necessary, therefore, to wash the cartridge only with diluted buffer after loading the sample.

After the sulfonamides were eluted from the SPE cartridge with 80% acetonitrile, it was not possible to inject the sample directly into the HPLC due to the amount of acetonitrile. Therefore, acetonitrile was evaporated from the sample using a rotatory evaporator before injecting the sample into the HPLC.

The method used by the FDA requires two different HPLC conditions to analyze for ten sulfonamides. For the FDA method, either two HPLC systems are required, or first seven sulfonamides are analyzed and then HPLC conditions are changed to analyze the remaining three sulfonamides. It was also necessary to clean the column after few runs in order to avoid interference from any later eluting peaks. In the method reported, the use of a gradient was found very convenient, time saving, and also increased resolution. All nine sulfonamides could be analyzed in 32 minutes and equilibration of the column back to initial conditions takes only about 20 - 30 minutes.

REFERENCES

1. Federal Register, 53 FR 9492, March 23, 1988, National Center for Toxicological Research Technical Report Experiment Number 418, March 1988, NCTR Jefferson, AR 72079.

2. A. R. Long, L. C. Hsieh, M. S. Malbrough, C. R. Short, and S. A. Barker, Multiresidue Method for the Determination of Sulfonamides in Pork Tissue, J. Agri. Food Chem., 38:423 (1990).

3. J. D. Weber and M. D. Smedley, Unpublished Milk Survey, Food and Drug Administration, Washington, DC (1988).

4. L. Larocque, G. Carignan, and S. Sved, Sulfamethazine (Sulfadimidine) Residues in Canadian Consumer Milk, J. Assoc. Off. Anal. Chem., 73:365 (1990)

5. Code of Federal Regulation, Title 21, 556.640, 473 (1990).

6. D. E. Dixon-Holland, and S. E. Katz, Direct Competitive Enzyme Linked Immunosorbent Assay for Sulfamethazine (Sulfadimidine) Residues in Milk, J. Assoc. Off. Anal. Chem., 72:447 (1989).

7. A. F. Lott, R. Smither, and D. R. Vaughan, Antibiotic Identification by High Voltage Electrophoresis Bioautography, J. Assoc. Off. Anal. Chem., 68:1018 (1985).

8. C. W. Sigel, J. L. Woolley Jr., and C. A. Nichol, Specific TLC (thin layer chromatographic) Tissue Residue Determination of Sulfadiazine Following Fluorescamine Derivatization, J. Pharma. Sci., 64:973 (1975).

9. J. E. Roybal, S. B. Clark, J. A. Hurlbut, C. A. Geisler, R. J. Schmid, and S. L. Cross, A Rapid Thin Layer Chromatographic Screening Procedure for the Detection of Sulfamethazine in Milk, Laboratory Information Bulletin, 6:3433 (1990).

10. S. B. Clark, R. G. Burkepile, S. L. Cross, J. M. Storey, J. E.

Roybal, and C. A. Geisler, A Rapid Thin Layer Chromatographic Screening Procedure for the Detection of Eight Sulfomamides in Milk, Laboratory Information Bulletin, 7:3528 (1991).

11. H. Holtmannspotter and H. P. Thier, Determination of Residues of Six Sulfonamides and Chloramphenicol by Gas Chromatography on Glass Capillary Columns, Dtsch. Lebensm. Rundsch. 78:347 (1982).

12. M. A. Alawi and H. A. Ruessel, Determination of Sulphonamides in Milk by High Performance Liquid Chromatography with Electrochemical Detection, Fresenius'Z. Anal. Chem. 307:382 (1981).

13. R. Malisch, Multi-method for Determination of Residues of Chemotherapeutic and Antiparasitics and Growth Promoters in Foodstuffs of Animal Origin. I. General Procedure and Determination of Sulfonamides, Z. Lebensm. Unters. Forsch, 182:385 (1986).

14. M. M. L. Aerts, W. M. J. Beek, and V. A. T. Brinkman, Monitoring of Veterinary Drug Residues by a Combination of Continuous Flow Techniques and Column Switching High erformance Liquid Chromatography 1. Sulphonamides in Egg, Meat and Milk using Post Column Derivatization with Dimethylaminobenzaldehyde, J. Chromatography, 435:97 (1988).

15. J. D. Weber, and M. D. Smedley, Liquid Chromatographic Determination of Sulfamethazine in Milk, J. Assoc. Off. Anal. Chem. 72:445 (1989).

16. M. D. Smedley and J. D. Weber, Liquid Chromatographic Determination of Multiple Sulfonamide Residues in Bovine Milk, J. Assoc. Off. Anal. Chem. 73:875 (1990).

17. M. Petz, High Pressure Liquid Chromatographic Determination of Residual Chloramphenicol, Furazolidone and Five Sulphonamides in Eggs, Meat and Milk, Z. Lebensm. Unters. Forsch., 176:289 (1983).

18. J. Unruh, E. Piotrowski, and D. P. Schwartz, Solid Phase Extraction of Sulfamethazine in Milk with Quantitation at Low ppb Levels Using Thin Layer Chromatography, J. Chromatography, 519:179 (1990).

19. A. R. Long, C. R. Short, and S. A. Barker, Method for Isolation and Liquid Chromatographic Determination of Eight Sulfonamides in Milk, J. Chromatography, 502:87 (1990).

20. V. K. Agarwal, Detection of Sulfamethazine Residues in Milk by High Performance Liquid Chromatography, J. Liquid Chromatography, 13:3531 (1990).

21. K. Uekama, F. Hirayama, M. Otagiri, M. Otagiri, Y. Otagiri, and K. Ikeda, Inclusion Complexation of B- Cyclodextrin With Some Sulfonamides in Aqueous Solution, Chem. Pharm. Bull, 26:1162 (1978).

22. V. K. Agarwal, Solid Phase Extraction of Sulfonamides Using Cyclobond-I Cartridges, J. Liquid Chromatography, 14:699 (1991).

SULFADIMETHOXINE AND SULFAMONOMETHOXINE RESIDUE STUDIES IN

CHICKEN TISSUES AND EGGS

Tomoko Nagata, Masanobu Saeki, Tetsuya Ida and
Masayuki Waki

Public Health Laboratory of Chiba Prefecture
Nitona-cho, Chiba City, 280 Japan

INTRODUCTION

Sulfadimethoxine (SDM) and sulfamonomethoxine (SMM) are widely used, either alone or in combination with other antibacterials, as a growth promoter, or for prevention and treatment of infectious diseases (1,2).

Onodera et al. (3) have administered feeds containing 0.005% and 0.2% of SDM or SMM to chickens and determined the residues of SDM and SMM in tissues and eggs by the <u>Bratton & Marshall</u> method. In their study, SDM and SMM were not found either in the tissues or eggs of chickens which were administered 0.005% SDM or SMM in feeds. On the other hand, in the 0.2% SDM or SMM groups, after withdrawal, both drugs were found in yolk for 11 or 7 days, and in albumen for 5 days. However, the <u>Bratton & Marshall</u> method is not sensitive enough to detect sulfonamide residues in tissues and lacks specificity.

Terada et al. (4) continued oral administration with 20 mg/day of SDM and SMM to chickens for 4 days and found both residues by high performance liquid chromatography (HPLC) in tissues 1 day after withdrawal.

In our previous study (5), chickens were administered 100 mg/day of SDM and SMM for 7 days, and SDM was found in yolk and albumen 7 and 5 days after withdrawal, respectively.

However, detailed analytical data have not been published so far about SDM and SMM residues in chicken tissues and eggs in feeding experiments with low levels of SDM and SMM. From the standpoint of public health, residue studies of SDM and SMM were attempted. In the present study, broilers and laying hens were administered the feed containing 25, 50 and 100 ppm of SDM and SMM for 21 days, which are routine doses for preventing leukocytozoon, and fed with the drug free feeds thereafter. The residues of drugs in yolk and albumen were examined in the eggs laid by medicated Leghorn hens, and residues were determined in tissues of broiler chickens. During the experimental period, the residues of SDM and SMM in tissues and eggs were determined by HPLC.

Analysis of Antibiotic Drug Residues in Food Products of Animal Origin
Edited by V.K. Agarwal, Plenum Press, New York, 1992

EXPERIMENTAL

Feeding Schedule

   Test animals. (A) Laying hens - Twenty one white Leghorn hens, weighing
about 1,500 g at age 189 days, were equally divided into 7 groups of 3 hens. The
first group was fed with commercial feed during the experimental period as a
control. The second, third and fourth groups were administered the feeds
containing 25, 50 and 100 ppm of SDM and the fifth, sixth and seventh groups were
administered with feeds containing the same concentration of SMM, for 21 days and
thereafter, fed with the commercial feeds during the observation period. The
drug was given 24 hours before day 1 and withdrawn on day 21, 24 hours before day
1 of withdrawal. This experimental design is shown in Table 1. (B) Broilers -
One day old chicks were purchased and fed with the commercial starter feeds for
broiler chickens for 4 weeks. These chickens, including both sexes, were equally
divided into seven groups of 40 chickens. The first group was fed with the
commercial feed as a control. The second, third and fourth groups were
administered the feeds containing 25, 50 and 100 ppm SDM, and the fifth, sixth
and seventh groups were administered with the same concentrations of SMM in the
same way for 21 days, and thereafter, fed without drugs.

Table 1. Design of Feeding Experiment for Laying Hens.

Group Drug	No. of Hens	Administration (Days)				After Withdrawal (Days)				
		1	7	14	21	1	3	5	7	9
1 Control	3		**				**			
2 SDM	3		25 ppm				**			
3 SDM	3		50 ppm				**			
4 SDM	3		100 ppm				**			
5 SMM	3		25 ppm				**			
6 SMM	3		50 ppm				**			
7 SMM	3		100 ppm				**			

Sample: yolk, albumen.
 **   : commercial feeds only.
Day 1 : 24 hours after the start of administration or withdrawal.

   Feeds. (A) Laying hens - Three different amounts of SDM or SMM were added to
a small portion of standard feeds for laying hens, which had already been
filtered by seive, and mixed well. Then, each of these six kinds of mixtures
were gradually added to about 500 g filtered feeds while mixing. Using a feed
mixer, each of these feeds were further blended with the standard commercial
feeds for laying hens to make the final concentrations at 25, 50 and 100 ppm of
SDM or SMM. (B) Broilers - Three different amounts of SDM or SMM were added to a
small portion of cornstarch (Kanto Chemical Co. Inc., Tokyo, Japan), and mixed

well. Each of the above mixtures was gradually added to about 500 g cornstarch
while mixing. Using a feed mixer, each of these six kinds of cornstarch was
blended with the standard commercial feeds for broiler chickens to make the final
concentrations at 25, 50 and 100 ppm of SDM or SMM.

Samples. (A) Eggs - Throughout the experimental period, eggs were collected
every day, weighed, separated into yolk and albumen and stored at -20^{0}C until
analysis. (B) Tissues of Broilers - On day 3, 7, 14 and 21, from the beginning
of administration, 3 chickens were sampled from each group, blood from the wing
vein was collected with a syringe, and then the chicken was sacrificed by cutting
the jugular vein. After complete bleeding, liver, heart, spleen, gizzard, fat,
thigh muscles, and breast muscles were collected, separately. After withdrawal
of the drug, on days 1, 2, 5, 7, and 9, 3 chickens in each group were also
sacrificed in the same way as described above. These tissues collected were
stored at -20^{0}C until analysis. This experimental design is shown in Table 2.

Table 2. Design of Feeding Experiment for Broilers.

Group Drug	(ppm)	Administration (Days)				After Withdrawal (Days)				
		3	7	14	21	1	2	5	7	9
1 Control		3*	3	3	3	3	3	3	3	3
2 SDM	25	3	3	3	3	3	3	3	3	3
3 SDM	50	3	3	3	3	3	3	3	3	3
4 SDM	100	3	3	3	3	3	3	3	3	3
5 SMM	25	3	3	3	3	3	3	3	3	3
6 SMM	50	3	3	3	3	3	3	3	3	3
7 SMM	100	3	3	3	3	3	3	3	3	3

* : the number of sacrificed broilers.
Samples: blood, liver, heart, spleen, gizzard, fat, thigh muscle, breast muscle.
Day 1: 24 hours after the start of administration or withdrawal.

Analytical Method

Reagents. (A) Solvents - Acetonitrile, n-Hexane, n-propyl alcohol, acetic
acid (Wako Pure Chemical Industry LTD., Osaka, Japan). (B) Mobile phase for
column cleanup - Acetonitrile:water (85:15, v/v) (C) N-hexane saturated
acetonitrile - N-hexane was partitioned with acetonitrile before use. (D)
Alumina - Woelm B activity grade I (Woelm pharma GmbH & Co., FRG). Pack a
chromatographic column, 300 mm x 15 mm i.d., with 3 g of alumina suspended in
acetonitrile. Wash the column with 25 ml acetonitrile followed by 25 ml of (B)
solution. (E) LC Mobile Phase - Acetonitrile:acetic acid:water (30:1:70,
v/v/v). (F) Sulfadimethoxine (SDM) and Sulfamonomethoxine (SMM) standard
solution - Prepare stock solution of SDM or SMM at 100 ug/ml by dissolving 10 mg
SDM or SMM (Sigma Chemical Co., St. Louis. MO) in 100 ml of acetonitrile.

Prepare the working standard solution at 1 ug/ml in acetonitrile using the stock solution.

Apparatus. (A) Liquid chromatograph - JASCO Model 880-PU pump equipped with JASCO Model 875-UV ultraviolet spectrometer and JASCO Model 960-CO column oven (Japan Spectroscopic Co., LTD., Tokyo, Japan). Integrator: Shimadzu Model CR-6A (Shimadzu Sheisakusho Co., Kyoto, Japan). Chromatographic condition: flow rate, 1.0 ml/min.; temperature, 50^0C; detection, 270 nm. (B) Guard Column - Nucleosil C18 stainless steel, 50 mm x 4.6 mm i.d. (10 um, Gasukuro Kogyo, Inc., Tokyo, Japan). (C) Chromatographic column - Nucleosil C18 stainless steel, 250 mm x 4.6 mm i.d. (5 um, Gasukuro Kogyo, Inc.).

Extraction and cleanup. Accurately weigh 1-10 g of tissue, yolk or albumen, individually. Homogenize at moderate speed with 40 ml of acetonitrile, centrifuge at 3000 rpm for 10 minutes and transfer supernatant to a separatory funnel. Add 30 ml n-hexane (c), shake for 5 minutes and let stand until the two layers separate. Apply the lower phase on alumina column and drain solvent to ca 0.5 cm above alumina layer. Add 25 ml of (b) on the alumina column, elute SDM and SMM and collect all eluates. Add 20 ml n-propyl alcohol to the eluates and evaporate to dryness under vacuum on a rotary evaporator at 65^0C. Dissolve residues with 1 ml methanol in an ultrasonic bath, filter the solution through a filter membrane at 0.5 um porosity and put the filtrate into the LC instrument. In the case of thigh and breast muscles, yolk and albumen, the partition procedure of supernatant with n-hexane (c) is not necessary.

RESULTS

Analytical Studies

The standard curves of SDM and SMM were both linear over the range of 1-20 ng with the correlation coefficient of 0.9987 and 0.9998, respectively. The detection limit, defined as the 3 fold noise level, was 1 ng, which corresponded to 0.01 ppm SDM and SMM in tissues, yolk and albumen. Recovery studies were performed by adding 1 or 2 ml of the working standard solutions of SDM and SMM onto 10 g portions of each tissue, yolk and albumen. Recovery data on fortified tissues, yolk and albumen were calculated by comparing LC peak area with those of the working standards. The average recoveries from each tissue, yolk and albumen are shown in Table 3. Typical chromatograms of blank tissues, yolk and albumen without medication were shown in Figure 1.

Feeding Trial

(A) Laying hens. The average feeds intake, gain of body weight and egg production ratio in all 7 groups during the administration period were 97-109 g, 2.0-4.7 g and 62-86%, respectively. There was no significant difference among all 7 groups.

(B) Broilers. The average body weight of 4 week old chicks was 902 ± 17.5 g, and in all 7 groups during the administration period, the average of feed intake was 140-150 g/day, and gain of body weight was 41-63 g/day in females and 57-75 g/day in males. There was no significant difference among all 7 groups. In this study, no drug effect was observed either on the gain of body weight or on the egg production rate.

Concentration of SDM and SMM in Yolk and in Albumen During the Experimental Period

In the control group, SDM and SMM were not found either in yolk or albumen during the experimental period.

In the third group at 50 ppm SDM, SDM was first found in yolk on day two

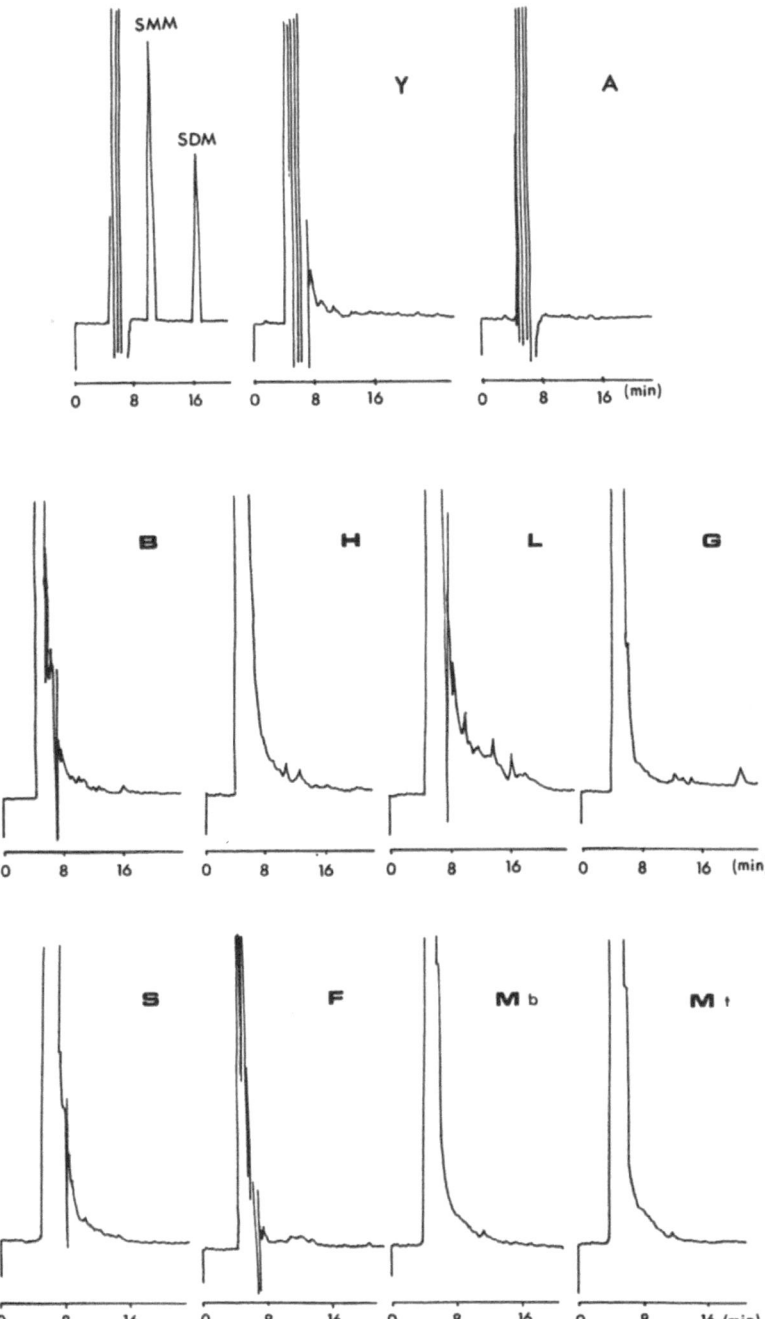

Figure 1. HPLC Chromatograms of SDM and SMM Standards and Chicken Tissue Extracts without Medication. SDM:20 ng, SMM:20 ng, Y:yolk, A:albumen, B:blood, H:heart, L:liver, G:gizzard, S:spleen, F:fat, Mb:breast muscle, Mt:thigh muscle.

177

from the beginning of administration and, thereafter, concentration of SDM increased gradually to reach the stable phase for a certain period seen as a loose plateau from day 14. The average concentration of SDM in the period from day 14 to 21, was 0.11 ppm. After withdrawal, the concentration of SDM in yolk decreased gradually to below the detection limit on day 7. In albumen, SDM was first found on the next day of the drug administration. However, the concentration of SDM in albumen varied day to day, therefore, no clear plateau was observed. The average concentration of SDM in albumen during the administration period was 0.24 ppm. After withdrawal, the concentration of SDM decreased below the detection limit on day 3.

Table 3. Average Recovery Percentage of SDM and SMM in Each Tissue.

Tissue	SDM		SMM	
	Added (ppm)	Recovery (%)	Added (ppm)	Recovery (%)
Yolk	0.2	90.3 (n=3)	0.2	92.2 (n=3)
Albumen	0.2	90.7 (n=3)	0.2	97.7 (n=3)
Blood	0.1	77.2 (n=3)	0.1	88.5 (n=5)
Heart	0.1	90.9 (n=3)	0.1	85.5 (n=5)
Liver	0.1	80.5 (n=3)	0.1	88.1 (n=5)
Spleen	0.1	73.0 (n=3)	0.1	84.6 (n=5)
Gizzard	0.1	91.3 (n=3)	0.1	85.1 (n=5)
Thigh Muscle	0.1	93.7 (n=3)	0.1	76.2 (n=5)
Breast Muscle	0.1	79.9 (n=3)	0.1	81.7 (n=5)
Fat	0.1	62.5 (n=3)	0.1	84.0 (n=5)

Even though drug concentrations in yolk and albumen were different in each medicated group, increase and decrease patterns of the drug in yolk and albumen were similar to those of the third group shown in Figure 2. These results mentioned above are summarized in Table 4.

Concentrations of SDM and SMM in Tissues During the Administration Period

In the control group, SDM and SMM were not found in any tissues during the experimental period. In the 6 medicated groups, SDM or SMM was found in almost all tissues from day 3 through day 21 during the administration period.

The minimum and maximum concentrations of SDM or SMM in each tissue on days 3, 7, 14 and 21 are shown in Table 5. In Figure 3, those of SDM in the third group are illustrated. Even though the drug concentrations were different, the distribution of drug concentration in the tissues of the other five medicated groups were similar to those in the third group.

Elimination of SDM and SMM From Tissues After Withdrawal

In the three groups medicated with SDM, 0.01-0.03 ppm of SDM was found in the liver 24 hours after withdrawal, as shown in Table 6. SDM was not found in tissues other than livers and one gizzard of the fourth group.

In the three groups medicated with SMM, 0.01-0.11 ppm SMM was found in half of the tissues. On day 2, SMM and SDM were not found in any tissues of any of the 6 medicated groups (data are not shown).

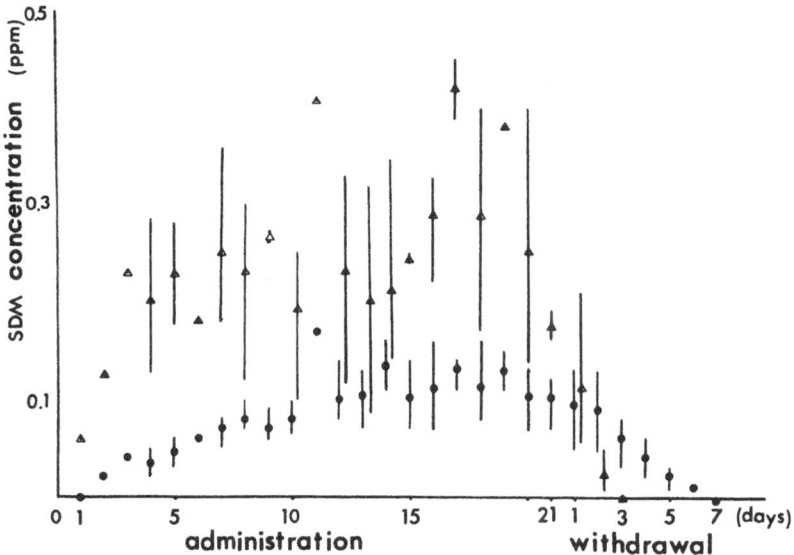

Figure 2. Concentrations of SDM in Yolk and Albumen of Eggs Laid by Hens Administered at 50 ppm of SDM in Feeds.

I : Range of minimum to maximum concentration of SDM
● : Mean concentration of SDM in yolk
△ : Mean concentration of SDM in albumen

DISCUSSION

Analytical Studies

Many HPLC methods have been used for determination of residual sulfonamides in tissues. The present method, modified in our previous study (6), is rapid, sensitive and suitable for the residue monitoring programs. The recoveries were satisfactory for monitoring residual SDM and SMM. In chromatograms of tissues, yolk and albumen, there were no extraneous interferring spikes around SDM and SMM peaks.

Table 4. Change of SDM and SMM Concentrations (ppm) in Yolk and Albumen by Drug Administration.

	Group			Administration Period		After Withdrawal
	Drug	(ppm)	1) day of first detection	2) plateau period (days)	3) avg. conc. at plateau	4) drug positive days
yolk	2 SDM	25	3	11-21	0.04	6
	3 SDM	50	2	14-21	0.11	6
	4 SDM	100	2	9-21	0.18	7
	5 SMM	25	3	5-21	0.04	4
	6 SMM	50	3	7-21	0.06	4
	7 SMM	100	2	10-21	0.17	7
albumen	2 SDM	25	2	12-21	0.12	2
	3 SDM	50	1	*5)	0.24	2
	4 SDM	100	1	6-21	0.46	4
	5 SMM	25	1	5-21	0.20	2
	6 SMM	50	1	*5)	0.28	2
	7 SMM	100	1	*5)	0.80	2

1) The first day when the drug was detected in yolk or albumen.
2) A period seen as a plateau.
3) The average concentration of drug in yolk or albumen in a plateau period.
4) Days while the drug residues were detected in yolk or albumen.
5) * : No plateau was recognized due to big daily fluctuations. The average concentration was calculated from a period for 21 days.

Table 5. Concentrations (ppm) of SDM and SMM in Tissues During the Administration Period.

Group Drug (ppm)				Blood	Heart	Liver	Spleen	Gizzard	Thigh Muscle	Breast Muscle	Fat
2	SDM	25	min.	0.16	0.10	0.13	0.08	0.07	0.05	0.06	0.02
			max.	0.31	0.18	0.21	0.10	0.13	0.10	0.09	0.06
3	SDM	50	min.	0.37	0.30	0.37	0.20	0.20	0.17	0.15	0.07
			max.	0.83	0.57	0.66	0.33	0.35	0.30	0.24	0.12
4	SDM	100	min.	0.65	0.37	0.50	0.24	0.32	0.21	0.22	0.07
			max.	1.43	0.75	0.90	0.48	0.56	0.40	0.41	0.29
5	SMM	25	min.	0.31	0.18	0.14	0.10	0.11	0.07	0.06	0.03
			max.	0.48	0.31	0.25	0.17	0.21	0.19	0.14	0.12
6	SMM	50	min.	0.58	0.34	0.25	0.21	0.23	0.16	0.15	0.05
			max.	1.62	1.03	0.70	0.54	0.62	0.43	0.47	0.25
7	SMM	100	min.	1.99	1.18	0.91	0.68	0.83	0.55	0.44	0.12
			max.	3.04	2.24	2.00	1.14	1.58	1.05	1.29	0.63

Table 6. SDM and SMM Concentrations (ppm) in Each Tissue 24 Hours After Withdrawal.

Group	Drug (ppm)		Blood	Heart	Liver	Spleen	Gizzard	Thigh Muscle	Breast Muscle	Fat
2	SDM	25 min.	nd	nd	nd	nd	nd	nd	nd	nd
		max.	nd	nd	0.02	nd	nd	nd	nd	nd
3	SDM	50 min.	nd	nd	nd	nd	nd	nd	nd	nd
		max.	nd	nd	0.03	nd	nd	nd	nd	nd
4	SDM	100 min.	nd	nd	0.01	nd	nd	nd	nd	nd
		max.	nd	nd	0.02	nd	0.01	nd	nd	nd
5	SMM	25 min.	0.01	nd	0.01	nd	nd	nd	nd	nd
		max.	0.02	0.01	0.02	0.02	0.01	nd	nd	nd
6	SMM	50 min.	nd	nd	0.01	nd	nd	nd	nd	nd
		max.	0.11	0.05	0.05	nd	0.04	0.02	0.02	nd
7	SMM	100 min.	0.06	0.03	0.05	nd	0.01	nd	nd	nd
		max.	0.06	0.04	0.07	0.01	0.02	0.01	0.02	0.01

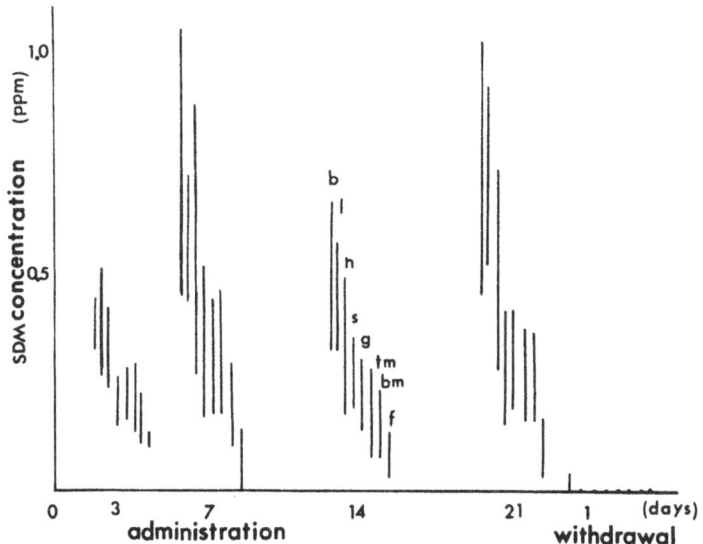

Figure 3. Concentration of SDM in Different Tissues of Broilers Administered 50 ppm of SDM in Feeds.

: Range of minimum to maximum average concentration of SDM
b:blood, l:liver, h:heart, s:spleen, g:gizzard,
tm:thigh muscle, bm:breast muscle, f:fat.

Feeding Trials

(A) Eggs. In this study, the routine doses for prevention of leukocytozoon infection, 25, 50 and 100 ppm of SDM and SMM in feeds were administered. After withdrawal, SDM remained in yolk for 6 to 7 days and in albumen for 2 to 4 days, and SMM remained in yolk for 4 to 7 days and in albumen for 2 days. The longer residue period of SDM and SMM in yolk has been observed in previous studies (3,4,5). This could be interpreted that since about 10 days are required for the growth of egg yolk in the ovary, some follicles, to which the drug had already been transferred in the medication period, are retained in the ovary for several days (7). In these studies, the drug concentration of SDM or SMM in albumen varied day by day while hens were fed with medicated feeds. However, with forced oral administration of the drug; as in the results of Terada et al. (4), and in our previous study (5), the concentration of drug in albumen does not vary so markedly day by day. In feeding experiments, the drug concentration in albumen seems to reflect the daily variation of medicated feed intake. Onodera et al. (3) described that neither SDM nor SMM residue was found in egg at 50 ppm SDM or SMM in feeds in his study. These results, different from ours, are attributed to the different sensitivities or detection limits; 0.01 ppm for our method, and 1.0 ppm for theirs.

As shown in Table 7, the proportion of SDM concentration in yolk to that in albumen, in the second, third and fourth groups, during the administration period was 0.33 to 0.46. That proportion was 0.5 in our previous study (5), and 0.35 in the study of Onodera et al. (3). It is concluded in agreement with previous investigations that SDM was mainly transferred to albumen rather than to yolk. These phenomena may be interpreted to indicate that sulfonamide had been hardly transferred to follicle tissue resulting in the low concentration of sulfonamide in yolk. In addition to the effect referred to above, sulfonamides increase their solubility in basic solution and they would be more readily transferred to albumen (pH 9.0), than to yolk (pH 6.5) (7). As shown in Table 7, the proportion of SMM concentration in yolk to that in albumen was about 0.21. In this experiment, SMM would be more easily transferred to albumen than SDM. Since the protein binding ratio of SDM is higher than that of SMM (7), SDM is presumed to be hardly transferred from blood to albumen.

As shown in Table 7, the transfer ratio of SDM from feeds to egg was 0.16 to 0.17%. This ratio is one third of the result of the previous study (5) by forced oral administration of the drug. The transfer ratio of SMM from feeds to egg was 0.19 to 0.29%. The transfer ratios of SMM and SDM in these feeding levels were nearly the same and not proportional to the amount of drug intake. As for the sixth group at 50 ppm SMM, the concentrations of SMM in yolk and albumen were comparatively low. As the feed intake of this group was similar to that of the other groups, SMM concentration in feeds was fell short of 50 ppm by mistake. The transfer ratio of SMM was about two times as much as that of SDM. This could be caused by the observation that the concentration of SMM in albumen was nearly two times as much as that of SDM.

(B) Broilers. In groups at 25, 50 and 100 ppm SDM, a determination of statistical significance was carried out by a F-test on each tissue on the average concentrations of SDM among days 3, 7, 14 and 21. There was no significant difference among the four days in every tissue. SDM was observed to be transferred into each tissue quite rapidly from the beginning of administration.

In groups of SMM, the significance test was also carried out in the same way. Only in the group at 50 ppm SMM, was a significant difference among 4 days (p<0.01) determined in most tissues, because of the remarkable low concentration on day 7. Although a significant difference was noticed at 50 ppm of SMM, this would be of little meaning, since there was no significant difference among days in all tissues for the 25 and 100 ppm groups.

Table 7. The Ratio of Transferred SDM and SMM in Eggs to the Amount of Intake During the Administration Period.

Group Drug (ppm)			Feed Intake (g/day/hen)	Drug Intake[1] (mg/day/hen)	Concn.[2] (ppm)	Weight[3] (g)	Transfer[4] Ratio (%)
2	SDM	25	y	109.2	2.73	0.04	13.50	0.159
			a			0.12	31.82	
3	SDM	50	y	104.8	5.24	0.11	13.39	0.171
			a			0.24	31.14	
4	SDM	100	y	106.2	10.62	0.18	14.42	0.169
			a			0.46	33.35	
5	SMM	25	y	98.5	2.46	0.04	13.12	0.293
			a			0.20	33.52	
6	SMM	50	y	99.6	4.98	0.06	13.28	0.186
			a			0.28	30.24	
7	SMM	100	y	97.4	9.74	0.17	12.68	0.274
			a			0.80	30.74	

y: yolk, a:albumen

4) The drug transfer ratio (R):

$$R (\%) = \frac{(Cy \times Wy + Ca \times Wa)}{D} \times \frac{100}{1000}$$

where Cy and Ca = average concentrations[2] of drug in yolk or albumen, respectively.

Wy and Wa = average weight[3] of yolk or albumen, respectively.

D = average drug intake[1].

There was a tendency for higher drug residues with increasing medicated levels. However, there were considerable variations in the concentrations of drugs in tissues by individual chickens in the same group. These variations would be hard to control because it probably reflects the inconsistency of the eating habits in individual chickens (7), resulting in the difference mentioned above.

In this experiment, comparatively high concentrations of residual SDM and SMM were found in blood followed by liver or heart. On the contrary, lower concentrations were found in muscles and fat. These results agreed with that of Onodera et al. (3), who administered a large amount of drugs, a 20-80 fold concentration of SDM or SMM of the present experiment. They observed high concentrations of SDM and SMM, i.e., 59 and 40 ppm in plasma, 28 and 21 ppm in heart, 38 and 16 ppm in liver, respectively. However, Terada et al. (4) fed only a 1.3 fold concentration of SDM and SMM from this experiment and obtained higher concentrations of residual SDM and SMM, i.e, 39.9 and 49.2 ppm in blood, 6.56 and 5.31 ppm in heart and 6.22 and 4.47 ppm in liver, respectively. The reasons why the higher concentrations of residual drugs were observed, would be presumed as follows:
1) As they administered SDM and SMM with two other kinds of sulfonamides at the same time, the interaction of drugs must be considered, 2) As they used only 2 chickens in a group, the individual difference was large, and 3) As two chickens were sacrificed only three hours after administration of drugs in pellets by forced administration, a large amount of SDM and SMM would remain in the blood.

As for the elimination of SDM and SMM from tissues, Terada et al. (4) also described in their study that SDM and SMM decreased from tissues rapidly 24 hours after withdrawal and were not found in any tissues on day 3.

In the present study, 24 hours after withdrawal, SMM remained more in tissues than SDM. This result could be caused by the fact that SMM concentration in tissues was 2 fold that of SDM during the administration period.

Concentrations of SDM and SMM in yolk were close to the levels in muscles and fat, and those in albumen were close to those in spleen and gizzard.

SDM and SMM were found to be transferred to tissues and albumen rapidly, and after withdrawal eliminated more rapidly from tissues than albumen, especially from yolk.

In conclusion, in our feeding experiments, the drug was first detected very early, mostly from the next day in tissues and albumen, somewhat later in yolk on days 2 or 3. The drug disappeared early from tissues within 48 hours, and from albumen mostly within 2 days. In yolk, the drug was, however, detected a little later, 2 or 3 days after start of feeding and remained a little longer, 4 to 7 days, after withdrawal.

ACKNOWLEDGMENTS

The authors are grateful to Dr. Vipin Agarwal to invite the present study to this symposium, and to Professor Hiroshi Tanaka, M.D., the Director of this Institution for his critical review and helpful suggestions on the manuscript preparation.

REFERENCES

1. R. Ewing, "Poultry Nutrition" 5th, The Ray Ewing Company, California (1963).

2. E. Takabatake, Feed additives and drugs for animal use, Eisei Kagaku, 27:127 (1981).

3. T. Onodera, S. Inoue, A. Kasahara and Y. Oshima, Experimental studies on sulfadimethoxine in fowls, Jap. J. Vet. Sci., 32:275 (1970).

4. H. Terada, M. Asanoma, H. Tubauchi, T. Ishihara and Y. Sakabe, Studies on residues and depletions of sulfonamides in chicken, Annual Report of Nagoya City Health Research Institute, 30:42 (1984).

5. T. Nagata, M. Masanobu and K. Toriumi, Transfer of pyrimethamine, sulfadimethoxine and difurazone into the eggs by oral administrations, Annual Report of Chiba Prefecture Institute of Public Health, 12:45 (1988).

6. T. Nagata and M. Saeki, Simultaneous determinations of 17 antibacterials in chicken tissues by high performance liquid chromatography, J. Food Hyg. Soc. Japan, 29:13 (1988).

7. I. Tazaki, I. Yamada, T. Morita and K. Tanaka, "Handbook of Poultry" (Japanese text), Youkendou Company, Tokyo (1985).

THE CURRENT OVERVIEW OF FEED ADDITIVES

AND VETERINARY DRUGS AND THEIR RESIDUAL

ANALYSIS IN JAPAN

Hiroyuki Nakazawa[1], Masahiko Fujita[1], Masakazu Horie[2]

[1]Department of Pharmaceutical Sciences, National
Institute of Public Health, Tokyo, Japan
[2]Saitama Prefectural Institute of Public Health,
Saitama, Japan

INTRODUCTION

Recently, the average life span of Japanese male and female has been the highest in the world. Although several reasons are considered and pointed out, the improvement of their eating habits might highly contribute to the increased life span. Since the latest Japanese, particularly young people, prefer Western style meals, which are mainly composed of meat, to representative Japanese style, the livestock industry has grown prosperous. The breeding scale in the livestock industry has been enlarged and become intensive year by year in order to reduce running expenses. As shown in Fig. 1, broilers or poultry industry has taken the large-scale breeding.

The populations of cattle and swine are similar to that described above. In Japan, the style of fisheries has been changed, owing to 200 seamile regulation, changes of preference of people for fishes, technological development and so on. The production of marine culture industry in Japan has increased during last 20 years.

In the raising industries for livestock and fisheries, a large population for breeding lives in a small area, so occurence of disease has increased and if it happens, the event causes a heavy loss. So, many kinds of antibiotics and synthetic antibacterials have been used in large quantities in order to prevent and treat the disease. On the other hand, feed additives are also used as an effective means to decrease economically the cost of production, and to improve the quality of product of livestock and fisheries and raise the productivity.

FEED ADDITIVES AND VETERINARY DRUGS USED IN JAPAN

To prevent the disease, antibiotic preparations and synthetic antibacterial agents, as well as feed additives, have become indispensable for production in livestock and fisheries industries.

Analysis of Antibiotic Drug Residues in Food Products of Animal Origin
Edited by V.K. Agarwal, Plenum Press, New York, 1992

187

As feed additives, many substances are used for three main purposes, that is prevention of deterioration of feed quality, supplementation of nutrients, and promotion of efficient use of animal feed ingredients. Anti-oxidants, antiseptics, emulsifiers, etc, have been used for the prevention of deterioration of feed quality due to growth of fungi and other causes, and amino acids, vitamins, and minerals have been used for promotion of growth. In addition, synthetic antibacterials and antibiotics have been used in order to promote the growth of animals at growing period and for prevention of infectious disease. These substances are specified in the Law concerning Safety Assurance and Quality Improvement of Feed (Government Ordinance No. 68 of 1976).

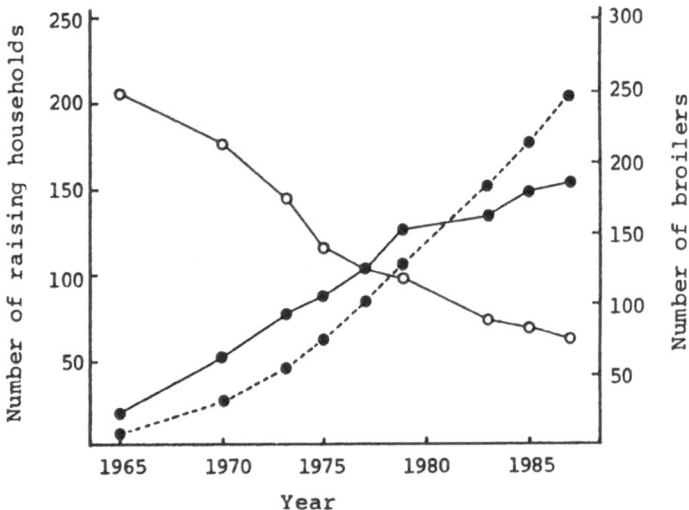

Fig. 1. Comparison of annual population of broilers and number of raising households.
- o——o: Number of raising households (x 100)
- ●——●: Total number of broilers (x 1,000,000)
- ●·····●: Number of broilers per one farm (x100)

On the other hand, synthetic antibacterials and antibiotics are used as veterinary drugs for the prevention and treatment of infectious diseases with relatively large dosage for short prescription periods. Current regulated synthetic antibacterials as feed additives and veterinary drugs by the Pharmaceutical Affairs Law and the law concerning Safety Assurance and Quality Improvement of Feed in animal husbandry and aquaculture are summarized in Table 1. A summarized list of synthetic antibacterials as veterinary drugs and feed additives is given in Table 2.

LEGAL RESTRICTIONS OF RESIDUES IN FOODS

In recent years, public concern over the presence of drug residues in meat products has rapidly grown in Japan. To prevent the residues of veterinary drugs, the law prescribes that animals should not be slaughtered shortly after the drugs are administered and while the concentration of the drugs remains at therapeutically effective levels. However, the illegal use or overdosed use of these compounds is occasionally found.

According to the Japanese Food Sanitation Law, no food should contain antibiotics and, in addition, meat, poultry eggs, fish and shellfish should not contain any synthetic antibacterial substances.

Table 2. Synthetic Antibacterials as Veterinary Drugs and/or Feed Additives

Synthetic Antibacterials		Veterinary Drugs[a]	Feed Additives[b]
Sulfonamides	Sulfadiazine	*	
	Sulfadimethoxine	*	
	Sulfadimidine	*	
	Sulfadoxine	*	
	Sulfisomidine	*	
	Sulfisoxazole	*	
	Sulfamethoxazole	*	
	Sulfamethoxypyridazine	*	
	Sulfamonomethoxine	*	
	Sulfaquinoxaline	*	*
	Sulfathiazole	*	
	Sulfachlorpyridazine	*	
	Sulfamoyldapsone	*	
Furan deriv-atives	Difurazon	*	
	Furazolidone	*	
	Nitrofurazone	*	
	Nifurstyrenic acid	*	
Antiprotozoan agents	Amprolium		*
	Clopidol		*
	Decoquinate		*
	Ethopabate		*
	Nicarbazin		*
	Pyrimethamine	*	
Others	Carbadox	*	
	Morantel citrate		*
	Nalidixic acid	*	
	Olaquindox		*
	Ormetoprim	*	
	Oxolinic acid	*	
	Piromidic acid	*	
	Thiamphenicol	*	
	Trimethoprim	*	
	Calcium halofuginone polystyrenesulfonate		*

a) Veterinary drugs: regulated by the Pharmaceutical Affairs Law
b) Feed additives: regulated as feed additives in the law concerning Safety Assurance and Quality Improvement of Feed

Table 1. Antibiotics as Veterinary Drugs and/or Feed Additives

Antibiotics		Veterinary Drugs[a]	Feed Additives[b]
β-lactams	Amoxicillin	*	
	Ampicillin	*	
	Cloxacillin	*	
	Dicloxacillin	*	
	Mecillinam	*	
	Nafcillin	*	
	Penicillin G	*	
	Cepharonium	*	
	Cephazolin	*	
Aminoglycosides	Apramycin	*	
	Destomycin A	*	*
	Dihydrostreptomycin	*	
	Fradiomycin	*	
	Gentamycin	*	
	Hygromycin B	*	*
	Kanamycin	*	
	Kasugamycin	*	
	Spectinomycin	*	
	Streptomycin	*	
Tetracyclines	Chlortetracycline	*	*
	Doxycycline	*	
	Oxytetracycline	*	*
	Tetracycline	*	
Macrolides	Carbomycin	*	
	Erythromycin	*	
	Josamycin	*	
	Kitasamycin	*	*
	Oleandmycin	*	*
	Sedecamycin	*	
	Spiramycin	*	*
	Tylosin	*	*
Polypeptides	Avoparcin		*
	Bacitracin	*	*
	Colistin	*	*
	Enramycin	*	*
	Mikamycin	*	
	Polymixin B	*	
	Thiopeptin	*	*
	Virginiamycin	*	*
Polysaccharides	Flavophospholipol	*	*
	Macarbomycin	*	
	Quebemycin	*	
Polyethers	Lasalosid		*
	Monensin	*	*
	Salinomycin	*	*
Others	Bicozamycin	*	*
	Chloramphenicol	*	
	Fosfomycin	*	
	Lincomycin	*	
	Nosiheptide		*
	Nystatin	*	
	Novobiocin	*	
	Tiamulin	*	

a) Veterinary drugs: regulated by the Pharmaceutical Affairs Law
b) Feed additives: regualted as feed additives in the law
   concerning Safety Assurance and Quality Improvement of Feed

Therefore, a simple and reliable method is required to monitor drug residues in edible tissues of swine, cattle and chicken. Monitoring samples obtained at abattoirs or collected by food inspectors are analyzed by the laboratories of government authorities. The development of analytical methods have made it possible to detect the residual drugs at trace levels. The necessary detection limit of each drug in livestock tissue should be defined by the evaluation of safety and toxicological aspects. So far, the level of 50 ppb has been considered as provisional dectection limits for analysis of most drugs.

RESIDUAL ANALYSIS OF VETERINARY DRUGS

The purpose of a screening test is to give a quick result whether the analyte is either not present in the sample or is below the level of concern. As the bioassay for detection of the presence of drugs, microbial inhibition test organized by Veterinary Sanitation Division, the Ministry of Health and Welfare, Japan, use antibiotic-sensitive strains of bacteria as the test organisms, as shown in Table 3. The Disk Assay has been used for detecting antimicrobial drugs, particularly antibiotics residues in livestocks in Japan.[1] For the assay of residual synthetic antibacterials, various chemical techniques have been developed using spectrophotometric procedures, thin-layer chromatography, enzyme immunoassay, gas chromatography (GC), high performance liquid chromatography (HPLC) and mass spectrometry (MS). The Japanese Ministry of Health and Welfare also prescribed chemical assay for the residual synthetic antibacterials as shown in Table 4.[2] As stated in Table 4, HPLC methods using a UV detector have recently been applied for the routine residual analysis[3]. It is to be noted that GC techniques have been gradually replacing HPLC for the analysis of residual drugs due to their ease of manipulation. However, conventional HPLC methods lack qualitative information. It is necessary to confirm the identification of observed peak before taking regulatory action from the standpoint of food hygiene. Although a few GC-MS methods have been developed for confirmation of target compounds, these are complicated and time consuming to prepare the suitable volatile derivatives for the large scale sceening purpose. For this purpose, other instrumental analysis using photodiode-array detector[4] or mass spectrometry[5] are powerful ways to identify the compounds in comparison with authentic compound. Fig. 2 shows typical chromatogram of commercial pork samples in which sulfadimidine, widely used in the rearing of food-producing animals to prevent and treat diseases and to promote their growth, was detected at 0.1 μg/g[6]. The peak component with a retention time of 7.1 min was compared with a standard sample of sulfadimidine. The high similarity index representing the similarity of the two spectra shows that the two spectra were almost identical, confirming the peak as sulfadimidine.

The other confirmation way of target compound found in sample is combination with MS. Fig. 3 demonstrates the mass spectra of standard oxolinic acid ($C_{13}H_{21}NO_5$, MW 261), which is widely used in fish to treat a variety of gram-negative organisms, and oxolinic acid found in a sweet fish sample. MS identification was carried out to confirm oxolinic acid in fractions obtained preparatively by HPLC. The retention of the sample peaks coincided with those of the standard oxolinic acid. The mass spectrum of both sample and standard oxolinic acid contain a molecular ion peak at m/z 261 and a parent peak at m/z 217. Therefore, the compound found in sweet fish was identified as oxolinic acid. By application of this method to the analysis of 115 commercial fishes, including yellowtail, eel, common carp, rainbow trout and sweet fish, oxolinic acid was found in 24 samples of sweet fish at levels ranging

Table 3. Microbiological Assay for Residual Antibiotics in Livestock Products in Japan [a]

Test organisms	Antibiotics
■Individual methods	
Bacillus stearothermophilus var calidolactis C-953	Ampicillin, Cloxacillin Lasalosid, Dicloxacillin, Salinomycin
Bacillus cereus var. mycoides ATCC 11778	Oxytetracylcine, Chlortetracycline
Micrococcus flavus ATCC 10240	Bacitracin
Micrococcus luteus ATCC 9341	Tylosin, Spiramycin, Kitasamycin, Oleandmycin, Erythromycin, Penicillin G
Bacillus subtilis ATCC 6633	Kanamycin, Streptomycin, Enramycin Quebemycin Monensin
Bacillus cereus ATCC 19637	Flavophospholipol
Pseudomonas syringae X 205	Hygromycin B
Corynebacterium xerosis NCTC 9755	Thiopeptin, Virginiamycin
Bordetella bronchiseptica ATCC 4617	Colistin
Bacillus brevis ATCC 8185	Macarbomycin, Destomycin A
Staphylococcus epidermidis ATCC 12228	Fradiomycin, Novobiocin
Escherichia coli NIHG	Chloramphenicol
Piricularia oryzae	Kasugamycin
■ Systematic methods	
Bacillus cereus var. mycoides ATCC 11778	Tetracyclines (Oxytetracycline, Chlortetracycline)
Micrococcus luteus ATCC 9341	Macrolides (Tylosin, Spiramycin, Kitasamycin Oleandmycin)
Bacillus subtilis ATCC 6633	Aminoglycosides (Streptomycin, Kanamycin, Fradiomycin, Destomycin A, Hygromycin B)

a) Official Analytical Methods for Residual Substances in Livestock Products, Vol. 1, Veterinary Sanitation Division, Environmental Health Bureau, Ministry of Health and Welfare, Japan.

Table 4. Chemical Assay for Residual Synthetic Antibacterials in
Livestock Products in Japan [a]

Antibacterials	Extraction and/or deprotienization	Measurement	Detection limit (ppm)
Sulfonamides	Acetonitrile	ECD-GLC (5% OV-17)	0.01-0.05
Furazolidone	Ethyl acetate	ECD-GLC (5% EGSS-x)	0.03
Difurazon	Ethyl acetate	TLC-densitometry (420nm)	0.1
Nifulpirinol	Acetone	UV-HPLC(Gel[b], 360 nm)	0.2-0.4
Nifurstyrenic acid	Methanol	TLC-densitometry(400/520nm)	0.05
Pyrimethamine	Isobutanol-benzene	ECD-GLC (1.5% OV-17)	0.05
Robenidine	Ethyl acetate	ECD-GLC (3% OV-17)	0.05
Dinitolumid	Acetonitrile	ECD-GLC (1.5% OV-17)	0.01
Amprolium	Trichloroacetic acid	TLC-densitometry(400/460nm)	0.02
Decoquinate	Methanol-chloroform	Fluorimetry(270/380 nm)	0.1
Clopidol	Mehtanol	ECD-GLC (10% DC-200)	0.05-0.1
Nicarbazin	Acetonitrile	UV-HPLC(Gel[b], 340 nm)	0.03
Ethopabate	Acetonitrile	UV-HPLC (ODS, 270 nm)	0.02
Carbadox	Acetonitrile	UV-HPLC (ODS, 380 nm)	0.05
Olaquindox	Acetonitrile	UV-HPLC (ODS, 380 nm)	0.05
Thiamphenicol	Acetone	FPD-GLC (2% OV-17)	0.5
Ormetoprim	Ethyl acetate	UV-HPLC (ODS, 230 nm)	0.05
Trimethoprim	Ethyl acetate	UV-HPLC (ODS, 230 nm)	0.05
Morantel citrate	Dichloromethane	UV-HPLC (ODS, 320 nm)	0.05
Oxolinic acid	Dichloromethane	UV-HPLC (ODS, 254 nm)	0.05
Nalidixic acid	Dichloromethane	UV-HPLC (ODS, 254 nm)	0.05
Piromidic acid	Methanol-chloroform	UV-HPLC (ODS, 280 nm)	0.05

a) Official Analytical Methods for Residual Substances in Livestock
Products,
Vol. 2, Veterinary Sanitation Division, Environmental Health Bureau,
Ministry of Health and Welfare, Japan.
b) Styrene-divinylbenzene copolymer

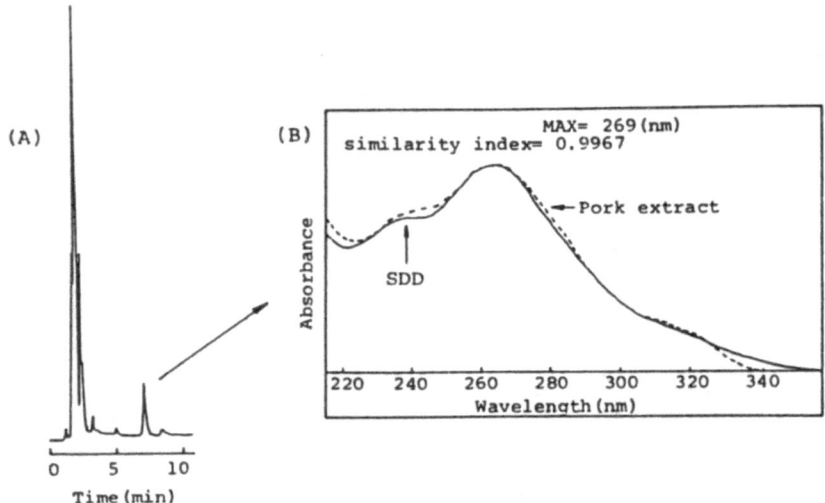

Fig. 2. (A) Chromatogram of pork sample in which sulfadimidine was detected at 0.1 µg/g, plotted at 275 nm. (B) Normalized spectra of the peak (at 7.1 min) obtained from pork extract (dashed line) and standard sulfadimidine (solid line). LC conditions: column, TSK-gel ODS 80T$_M$ (150 x 4.6 mm i.d.); mobile phase, 0.05M sodium dihydrogenphosphate-acetonitrile (2:1); flow rate, 0.5 ml/min; detector, Shimadzu SPD-M6A.

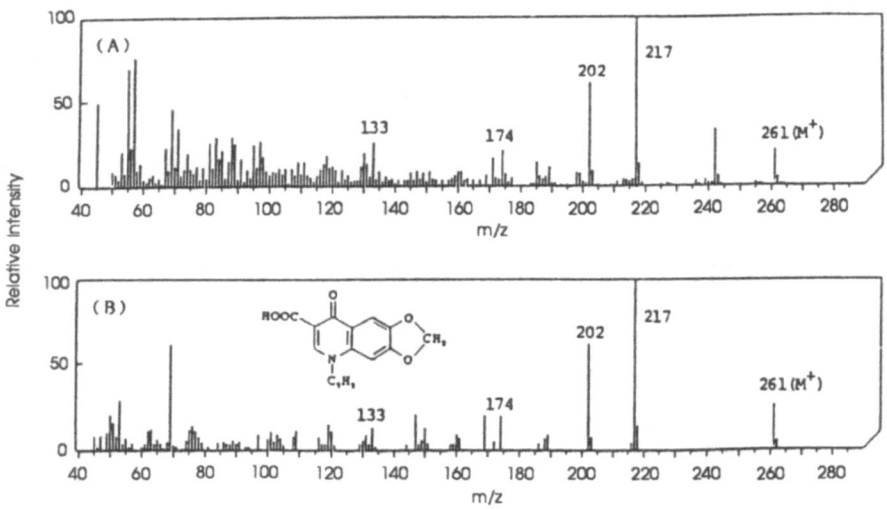

Fig. 3. Mass spectra of (A) oxolinic acid isolated by HPLC from sweet fish, and (B) authehtic drug.

from 0.01 to 1.90ug/g[7]. Furthermore, a thermospray HPLC-MS method has been developed for the analysis of sulfonamides, including sulfadimidine and sulfadimethoxine. The mass spectra obtained from sulfonamides were very simple with base peaks corresponding to protonated molecular ions MH[+]. An excellent correlation between the results of the thermospray HPLC-MS method and HPLC method was obtained (r=0.981). Recently, by this thermospray HPLC-MS, oxolinic acid in sweet fish was successfully confirmed with total ion chromatogram and mass chromatogram, as shown in Fig. 4. In addition to the above antibiotics, naturally occurring hormones such as progesterone and estradiol have been used for growth promotion and feed efficiency in heifers as anabolic agents. A vast assay of chemical techniques including radioimmunoassay, GC, GC-MS[8], and so on, have been developed for the determination of levels present[8]. Because of the trace amount of existing hormones, tedious and

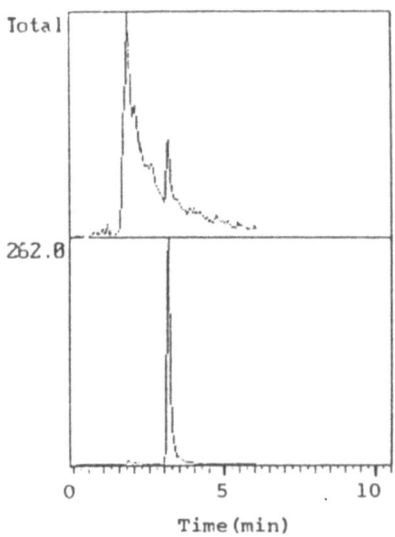

Fig. 4. Thermospray LC-MS total ion chromatogram and mass chromatogram of sweet fish sample containing 1.50 μg/g incurred oxolinic acid residue. Conditions: column, Inertsil ODS-2(150 x 4.6 mm i.d.); mobile phase, 0.05M ammonium acetate (pH 4.5)-acetonitrile (7:3); flow rate, 0.8 ml/min; column temperature, 35°C; vapour temperature, 165°C; ion source temperature, 270°C.

troublesome sample preparation steps, such as extraction and cleanup, are required for the analysis. Using the advantage of chromatography's separation ability, HPLC techniques also have been applied[9, 10]. However, the actual levels present are usually very low, and show wide variation, depending on the physical condition of the animals. Since these compounds are naturally present, all tissues and biological samples have these hormones as a constituent. Under present conditions, it is really hard to estimate the residual amount of exogenously administered anabolic agents.

SUMMARY

The current overview of feed additives and veterinary drugs and
their residual analysis in Japan has been reviewed. The problem of
drug residues in foods of animal origin has become increasingly
important to the entire livestock industry as growing consumer health
concerns. The development of new veterinary drugs must be evaluated
based on their efficacy, safety to the intended animal species and
safety to humans consuming products of animal origin. Moreover,
although residues of animal drugs in lifestock do not appear to be a
problem, it is necessary to survey the various products by appropriate
method. In future, evaluation of residues of veterinary drugs should
include the parent compounds and/or their metabolites in any edible
portion of the animal product. HPLC techniques with various detectors
can be expected to be successfully applied for the determiation of
residual feed additives and veterinary drugs, including anabolic agents,
in livestock.

REFERENCES

1. Official analytical methods for residual substances in livestock
   products, Vol. 1. Veterinary sanitation division, Environmental
   health bureau, Ministry of Health and Welfare, Japan.
2. Official analytical methods for residual substances in livestock
   products, vol. 2. Veterinary sanitation division, Environmental
   health bureau, Ministry of health and welfare, Japan.
3. H. Nakazawa and M. Fujita, Current overview of feed additives and
   veterinary drugs and their residual analysis, Eisei Kagaku.
   36:163(1990).
4. M. Horie, K. Saito, Y. Hoshino, N. Nose, N. Hamada and H. Nakazawa,
   Simultaneous determination of sulfa drugs in meat by high-
   performance liquid chromatography with photodiode-array detection,
   J. Food. Hyg. Soc. Jpn., 31:171(1990).
5. M. Horie, K. Saito, Y. Hoshino, N. Nose, M. Tera, T. Kitsuwa, H.
   Nakazawa and Y. Yamane, Simultaneous determination of sulfonamides
   in meat by thermospray liquid chromatography-mass spectrometry,
   Eisei Kagaku, 36:283 (1990).
6. M. Horie, K. Saito, Y. Hoshino, N. Nose, N. Hamada and H. Nakazawa,
   Identification and detrmination of suylphamethazine and N^4-
   acetylsulphamethazine in meat by high-performance liquid
   chromatography with photodiode array detection, J. Chromatogr.,
   502:371(1990).
7. M. Horie, K. Saito, Y. Hoshino, N. Nose, E. Mochizuki and H.
   Nakazawa, Simultaneous determination of nalidixic acid, oxolinic
   acid and piromidic acid in fish by high-performance liquid
   chromatography with fluorescence and UV detection, J. Chromatogr.,
   402:301(1987).
8. H. Nakazawa, Anabolic agents in meat production and their residue
   analysis, Bunseki No. 10:847(1989).
9. T. Miyazaki, T. Hashimoto, T. Maruyama, M. Miyazaki and H. Nakazawa,
   Determination of anabolic agents in beef by high operformance
   liquid chromatography, J. Food Hyg. Soc. Jpn., 30:384 (1989).
10. K. Watabe, H. Kikawa, T. Kawamura, T. Miyazaki, M. Matsumoto, H.
    Nakazawa, and M. Fujita, Simultaneous determination of zeranol, 17
    β-estradiol and diethylstibestrol in beef by HPLC with
    amperometric detection using column switching, Bunseki Kagaku,
    38:712(1989).

APPLICATION OF ELECTROCHEMICAL AND UV/VISIBLE DETECTION TO THE LC SEPARATION AND DETERMINATION OF METHYLENE BLUE AND ITS DEMETHYLATED METABOLITES FROM MILK

José E. Roybal, Robert K. Munns, David C. Holland, Jeffrey A. Hurlbut, and Austin R. Long

Food and Drug Administration, Animal Drug Research Center
Denver Federal Center, P.O. Box 25087
Denver, CO 80225-0087

ABSTRACT

A study was initiated to find a suitable HPLC detection scheme for the determination of methylene blue(MB) in several types of complex animal tissues, milk and milk product matrices. In addition to methylene blue, the detection system would be required to simultaneously detect several of its demethylated metabolites(DMM). Current HPLC methods involve detection in the visible absorbance range(600-660nm). Working in the visible range, however, does not provide for the maximum sensitivity of each component with a single injection. To achieve this, for the analytes of interest, one must have the ability to monitor each visible absorbance maxima; MB:661nm, azure A:633nm, azure B:648nm, azure C:616nm, thionin:598nm and toluidine blue O:626nm. Electrochemical detection(ED) is known to be one of the most sensitive detectors for HPLC. The basis for the chromatographic system used in this study is an isocratic elution of MB and its DMM which is performed on a cyano column with acetonitrile-acetate buffer as mobile phase. The order of elution, under these conditions, is thionin, azure C, azure A, toluidine blue O, azure B and MB, which corresponds to the degree of methylation of the thiazine structure(Figure 1). Presented is the evaluation and practicality of applying and utilizing electrochemical(ED) and/or UV/VIS detection for the analysis of MB and its DMM in milk.

INTRODUCTION

Methylene blue, 3,7-Bis(dimethylamino)phenothiazin-5-ium chloride, also known as C.I. Basic blue and tetrathionine chloride is widely used in bacteriology as a stain, as an oxidation-reduction indicator and as a reagent for several chemicals. Its primary use in the veterinary area is as an antiseptic, disinfectant and as an antidote for cyanide and nitrate poisonings (antimethemoglobinemic - a brownish-red crystalline, organic compound formed by the oxidation of hemoglobin and found in the blood after poisoning by chlorates, nitrates, ferricyanides or various other substances) including alcohol poisoning.

Several methods have been reported for the determination of methylene blue using HPLC or TLC(1-4,11-12,18-19). All published data, except Dean, et al(19), used its visible absorption spectra as the means of detection.

Analysis of Antibiotic Drug Residues in Food Products of Animal Origin
Edited by V.K. Agarwal, Plenum Press, New York, 1992

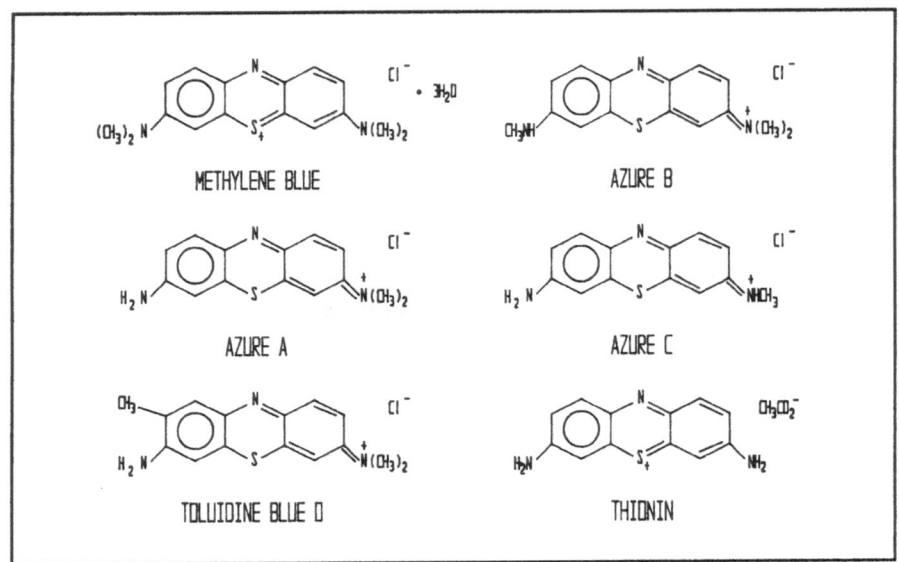

FIGURE 1.   CHEMICAL STRUCTURE OF ANALYTES OF INTEREST

Demethylation of methylene blue to N-methyl homologs of thionin has been documented(4,11).

DiSanto and Wagner(4) stated "early investigators did indicate that methylene blue was reduced in vivo and eliminated in unchanged form as well as in its leuco-form and as one or more 'chromogenic' substances in urine". They report that the leuco-form seemed to be stabilized by either salt, complex or combination in human urine but not in blood and tissue. Munns(18) found no evidence of leucomethylene blue in milk but that some form of a MB complex appeared to present.

Metabolic studies indicated the presence of the leuco-form (colorless) of at least methylene blue (4-8). DiSanto(4) reported the leucomethylene blue indirectly by acid hydrolysis.   Ziv, in 1984, proposed that the permeability of the blood-milk barrier to MB was based on the reduction to leucomethylene blue.In addition, the possibility remains for each of the demethylated metabolites to have a leucobase form. Figure 2 illustrates the probable metabolic conversion of MB based on the available information, leading to the wide variety of metabolites. Although Munns(18) could not verify that the leuco-form of MB was present in milk, the possibility that some metabolites of MB or its DMM might be overlooked monitoring the HPLC eluate at 627nm, had to be taken into account.

Little is known about the human food safety of methylene blue. Japanese investigators (9), in 1976, found methylene blue and thionin, as well as several other dyes, to show positive mutagenicity with Escherichia coli and Salmonella typhimurium strains.

Concern over the possible mutagenicity of MB and lack of adequate methodology made analytical procedures to identify and detect MB and all possible metabolites in milk and edible animal tissue a necessity.

Toluidine blue O was included in this investigation because it is often used in place of thionin and methylene blue as a nuclear stain.

In 1989, we published a high performance liquid chromatographic system utilizing an electrochemical detector(ED) to determine gentian violet(GV), its demethylated metabolites(DMM), leucogentian violet(LGV) and methylene blue(MB)(15). This system was successfully used to develop analytical procedures for the detection and determination of GV, its DMM and LGV in chicken tissue and fat at residue levels of <100ppb(16-17).

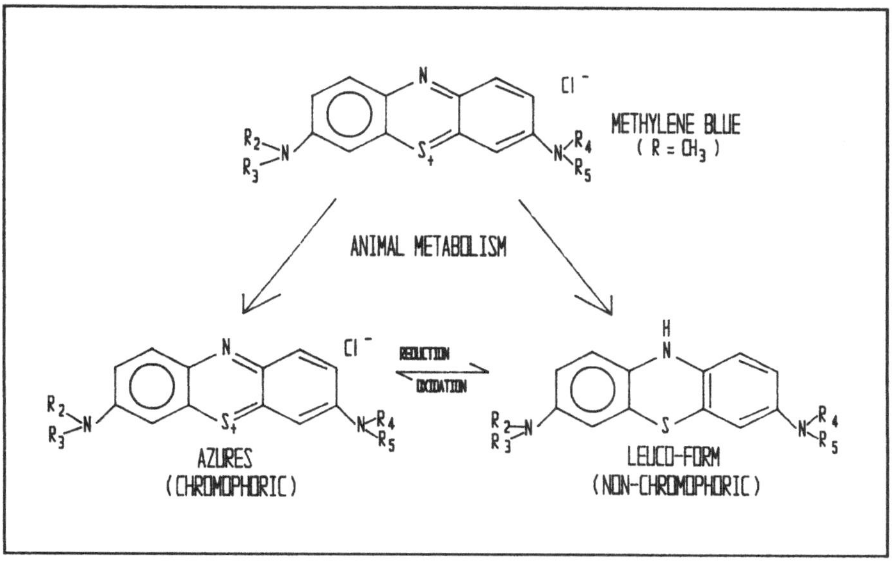

FIGURE 2. ANTICIPATED METABOLISM OF METHYLENE BLUE AND
ITS PROBABLE METABOLITES

Unlike GV and its DMM, where the UV/VIS maxima are very similar, 588nm, MB and its DMM have quite different UV/VIS maxima. Munns(18) recently reported a procedure using visible detection for the analysis of MB and its DMM in milk. Because they each have different visible absorption maxima, it was necessary to either make several injections of the sample extracts, varying the wavelength with each injections or finding a compromised wavelength to monitor all with a single injections but sacrificing sensitivity and specificity.

Our objective here was to evaluate an alternate detector to UV/VIS. A detector capable of monitoring MB and all possible metabolites (chromophoric/non-chromophoric) at ppb level. A complete chromatographic system applicable to various sample matrices.

METHOD

APPARATUS

(a) Syringes. - 10mL, B/D glass (Cat.# BD2312, Becton-Dickinson, Rutherford, NJ 07070).
(b) Pasteur pipet.- Disposable, 5.75 in.
(c) Centrifuge tube.- 15mL, graduated, with glass stopper #13 (Kimble cat. no. 45153-A) or equivalent.
(d) Centrifuge tube.- 150mL, Falcon Blue Max, disposable, graduated, conical, polypropylene with cap (Cat. No. 2076, Becton/Dickinson, Lincoln Park, NJ 07035) or equivalent.

(e) Liquid chromatograph .-LC pumping system.- Waters LC pump Model 6000-A and Model U6K universal LC injector (Waters Associates, Milford, MA 01757). Operating conditions: chart speed, 0.25 cm/min.; mobile phase flow, 1.0mL/min.; column temperature, ambient; column pressure, 2500 psi; volume injected, 20-50μL.

(f)Detectors. -(1) Electrochemical detector (ED).- Bioanalytical Systems Model LC-4B; working electrode, glassy carbon, single electrode vs Ag/AgCl ref. electrode; working potential, +1.000V; current range, 10nAFSD,(Bioanalytical Systems,Inc., Purdue Industrial Research Park, West Lafayette, IN 47906) (2) UV/VIS detector.- Shimadzu Ultraviolet-Visible spectrophotometric detector, SPD-6AV module for HPLC (Shimadzu Corporation, Analytical Instruments Division, Kyoto, Japan), cell volume, 8μL; light source, tungsten halogen (WI) lamp(370-700nm); wavelength, 627nm; absorbance range, 0.005 AUFS.

(g) LC column.- LC column.-Phenomenex Ultremex 5CN (cyano), 5 micron, 150mm x 4.6mm id. (Cat.# 00F-0050-EO [serial# PP/5675C], Phenomenex, Torrance,CA) or equivalent.

(h) Recorder.- Dual channel SE-120 strip chart recorder set at 10mV (BBC-Metrawatt/Goerz, Broomfield CO 80020).

(i) Filter - Millipore, disposable, 5 micron, PTFE membrane (cat. No. SLSR025NB) or equivalent.

(j)Rotary evaporator.- Buchi Model R-110 with ice trap (Brinkmann Instruments, Inc. Westbury, NY 11590) or equivalent.

REAGENTS

(a) Solvents.- Distilled-in-glass, pesticide-grade, methanol, and UV spectro-grade acetonitrile (Burdick & Jackson Laboratories, Inc., Muskegon, MI 49442) or equivalent.

(b) Water.- HPLC grade (Fisher Scientific) or deionized, glass-distilled.

(c) Acetic acid.- ACS grade, glacial, aldehyde-free.

(d) Sodium acetate.- ACS grade, anhydrous.

(e) EDTA. - Disodium Ethylenediamine Tetraacetate, ACS grade.

(f) Acetate buffer.- prepare by adjusting a 0.1M sodium acetate solution (8.2g sodium acetate/ 1000mL water), containing @ 50mg EDTA/L, to pH 4.5 with acetic acid (ca 8mL). Use to prepare mobile phase, (g).

(g) Mobile phase.- Acetonitrile/Acetate buffer (20+80).

(h) Reference Standards.- reference standard. Methylene blue,USP (Cat.# 42800, United States Pharmacopeial Convention,Inc., Rockville, MD 20852); Azure A and Azure B (Cat.# C8689 and C8683, respectively, Eastman Kodak Co., Rochester, NY 14650); Azure C, Thionin and Toluidine Blue O (Cat.# 24,218-7, 86,134-0 and 19,816-1 respectively, Aldrich Chemical Co., Milwaukee, WI 53233). Stock standards: (100ug/mL) Accurately weigh 10.0mg of each reference standard (adjust accordingly for label purity of each reference standard) into individual 100mL volumetric flasks, dilute to volume with methanol and mix. Intermediate Mixed standard: (2.0ug/mL) Pipet 2.0mL of each stock solution into 100mL volumetric flask, dilute to volume with methanol and mix. Working Mixed standard (MBMIX): (0.1ug/mL) Pipet 1.0mL of intermediate mixed standard into a 15mL centrifuge tube, dilute to 10.0mL with methanol and mix. Prepare weekly or as needed. Fortification Mixed standard: (10ug/mL) Pipet 1.0mL of each stock standard into 10mL volumetric flask, dilute to volume with methanol and mix. Aliquot 50, 100 and 200uL of this solution to 10mL of control milk for fortification levels of 50, 100 and 200 ppb.

SAMPLE ASSAY

Extraction: Aliquot 10mL of milk sample into a 150mL centrifuge tube(d). Add 75mL acetonitrile to sample, stopper and shake vigorously 30 seconds. Centrifuge sample at 2500rpm at 0° for 20 minutes. Decant CH_3CN supernatant into 150mL pear-shaped boiling flask. Roto-evaporate CH_3CN supernatant to approximately 5mL at 50°. To reduced extract, add 15mL methanol and roto-evaporate to just dryness. Dissolve residue in 5.0mL mobile phase. Filter sample extract through PTFE(teflon) filter. Inject 50uL filtered sample extract into liquid chromatograph. Bracket each sample set with 50uL injections of the mixed working standard solution.

RESULTS AND DISCUSSION

The analysis of milk for MB provides a unique and difficult challenge. The first consideration of this work was to establish proper chromatographic conditions for the separation of the analytes in question, Figure 1. The HPLC system used was a modification of the one reported earlier[15]. It consisted of using a cyanopropyl reverse phase column isocratically eluted with an acetonitrile/acetate buffer mobile phase. Performance parameters for this chromatographic system are shown in Table 1. The values show the excellent separating and resolving power of the chromatographic system for these dyes. EDTA was introduced into the mobile phase to prevent fouling of the working electrode by the oxidation of MB as reported by Munns[18].

Table 1.   Performance parameters for chromatographic system using 20:80 Acetonitrile/0.1M sodium acetate buffer, pH 4.5 + 50mg EDTA/L, Phenomenex 5CN, 150mm x 4.6mm column.

	k'	$\alpha(X_1/X_2)$	N	$H(10^{-4})$	$R(X_1/X_2)$
Thionin	2.00	-----	1296	1.16	-----
Azure C	3.33	1.67(AZC/Thio)	676	2.22	2.67(Thio/AZC)
Azure A	5.67	1.70(AZA/AZC)	1111	1.35	3.18(AZC/AZA)
Tol. B	8.00	1.41(TolB/AZA)	1296	1.16	2.59(AZA/TolB)
Azure B	10.3	1.29(AZB/TolB)	1156	1.30	2.00(TolB/AZB)
MB	18.0	1.74(MB/AZB)	1061	1.41	4.18(AZB/MB)

Where:

k'= capacity factor, ability of column to retain analyte.
$k' = V_1-V_0/V_0$

α = separation factor, column's ability to separate two components.
$\alpha = k'_2/k'_1$

N = theoretical plate number, empirical measure of column efficiency.
$N = 16(V_x/W_x)^2$

H = height equivalent to a theoretical plate, measure of column efficiency per unit length.
$H = L$(column length, m)/ N(theoretical plates number)

R = resolution, ability of column to separate adjacent peaks.
$R = (V_2-V_1)/\frac{1}{2}(W_2-W_1)$

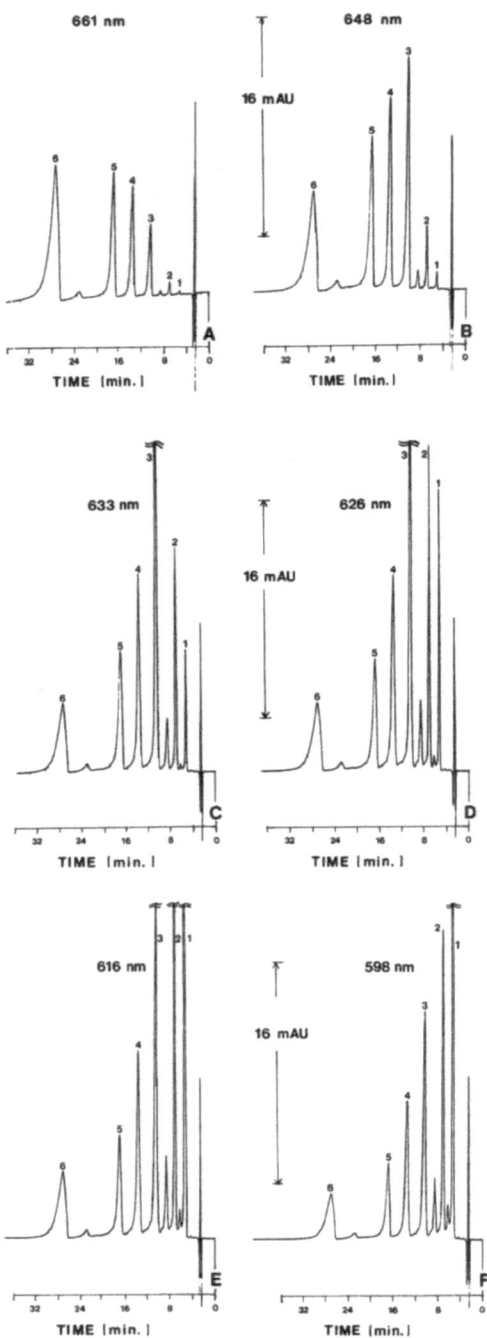

FIGURE 3. TYPICAL CHROMATOGRAMS OF INTERMEDIATE MIXED STANDARD. ORDER OF ELUTION, (1) THIONIN, (2) AZURE C, (3) AZURE A, (4) TOLUIDINE BLUE O, (5) AZURE B AND (6) METHYLENE BLUE. (A) @661nm, (B) @648nm, (C) @633nm, (D) @626nm, (E) @616nm AND (F) @598nm.

Figure 3 shows the response of the six analytes of interest (intermediate mixed standard) when monitored at each of their visible wavelength maxima; MB:661nm, azure A:633nm, azure B:648nm, azure C:616nm, thionin:598nm toluidine blue 0:626nm. While Figure 4 is the same analyte mixture monitored at their common UV wavelength maximum of 285nm and the ED of that same mixture.

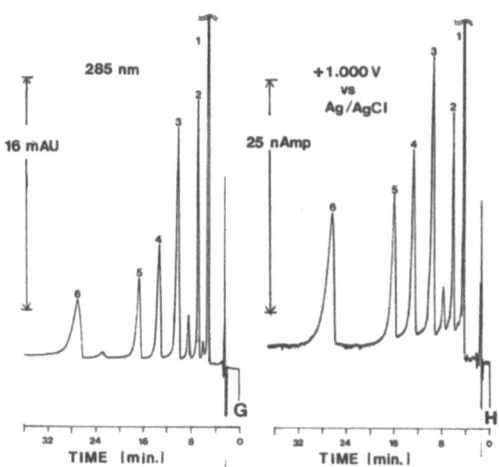

FIGURE 4.   TYPICAL CHROMAOTGRAM OF INTERMEDIATE MIXED STANDARD. (G) @285nm and (H) @ ELECTROCHEMICAL DETECTOR, +1.000V vs Ag/AgCl. ORDER OF ELUTION, (1) THIONIN, (2) AZURE C, (3) AZURE A, (4) TOLUIDINE BLUE O, (5) AZURE B AND (6) METHYLENE BLUE.

Figure 5 is the graphic representation of figures 3 and 4 (@ 285nm) showing the variation in response of each analyte. When working in the visible range a compromise wavelength must be decided upon which will detect all the compounds of interest. It is obvious that the response in the visible mode is severely scattered. Although all six have a common UV maximum of 285nm, working at this wavelength would have required a more extensive cleanup procedure. Our selection of 627nm as the wavelength for monitoring the HPLC eluate was based on the work by Munns(18). Also the data from figure 5 showed that at or near 626nm all the dyes gave a more comparable response. Our detector was optimized for toluidine blue O at 627nm.

A voltammogram of each of the six dyes was prepared using the Intermediate mixed standard solution, Figure 6. It is evident that the ED response of all six is more consistent and uniform then their visible absorption. From the voltammogram one can see that it would be possible to increase the sensitivity by using a working potential above 1000mV. In most cases, this offers no advantage in that response from other milk components would also become more pronounced. It would be equivalent to operating in the UV range at 200-220nm. While each does vary in the ED, their slopes are similar, leading to a more predictable and consistent response.

Typical chromatograms of control milk are depicted in Figure 7. Both detectors gave clean backgrounds for the milk matrix. Comparison of the detection of the 8hr post-dosing incurred milk is shown in figure 8 and 9, visible and ED, respectively. It is apparent that the determination of thionin is somewhat less precise with ED due to the fact of its early elution. Quantitation becomes less precise with peaks on the slope of the solvent/matrix front. Figure 10 and 11 are examples of the chromatography of 24hr post-dosing incurred milk as well as a 50ppb fortified sample, visible and ED, respectively.

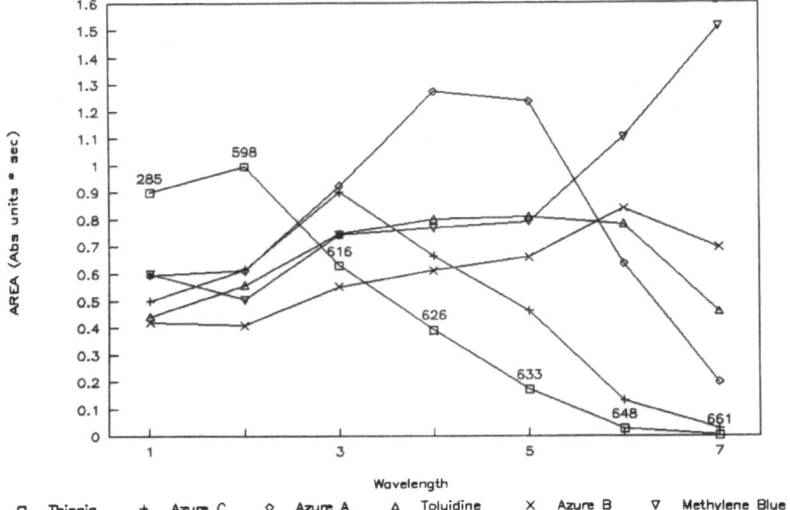

FIGURE 5.   GRAPHIC REPRESENTATION OF CHROMATOGRAMS IN THE UV/VIS. FIGURE 3 AND FIGURE 4 (@285nm).

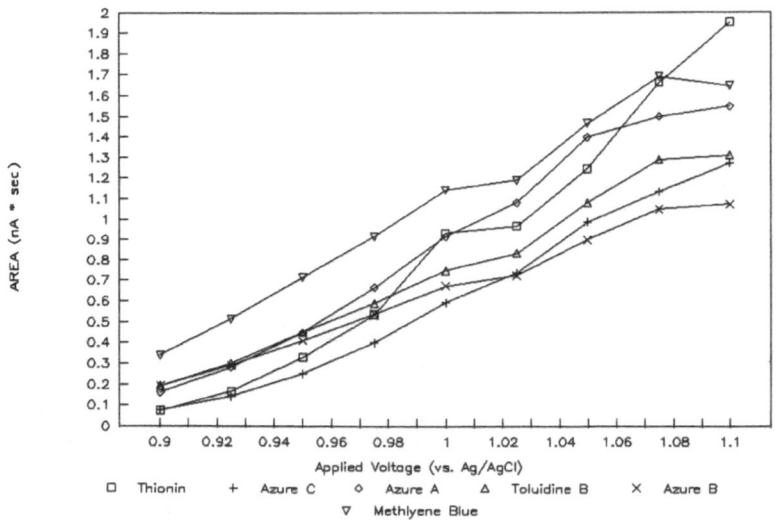

FIGURE 6.   VOLTAMMOGRAM OF THIONIN, AZURE C, AZURE A, TOLUIDINE BLUE O, AZURE B AND METHYLENE BLUE. APPLIED VOLTAGE RANGE vs Ag/AgCl REFERENCE ELECTRODE, +1.100V TO +0.900V IN 0.025V INCREMENTS.

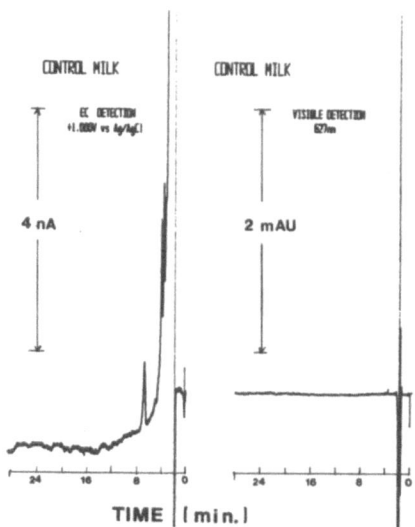

CONTROL MILK   CONTROL MILK

EC DETECTION   VISIBLE DETECTION
+1.000V vs Ag/AgCl   627nm

4 nA   2 mAU

24   16   8   0   24   16   8   0

TIME [min.]

FIGURE 7.   TYPICAL CHROMATOGRAM - CONTROL MILK, 10mL OF CONTROL MILK
             THROUGH METHOD. FINAL DILUTION IN 5.0mL MOBILE PHASE, 50uL
             INJECTION: A - VISIBLE DETECTION AT 627NM, 0.005 AUFS
             B - ELECTROCHEMICAL DETECTION AT +1.000V vs Ag/AgCl, 10nA FS

Table 2.   RECOVERY OF METHYLENE BLUE RESIDUES FROM FORTIFIED MILK

PERCENT RECOVERY

ELECTROCHEMICAL DETECTION, +1.000V vs Ag/AgCl

spike level ppb	THIO	AZ C	AZ A	TOL.B	AZ B	MB
50	n/c	56	83	77	92	100
"	n/c	60	92	71	98	100
"	9	12	53	37	59	62
"	16	48	65	62	70	72
"	12	56	77	69	82	81
"	14	48	81	70	89	89
100	16	70	90	82	98	96
"	10	67	96	60	102	88
200	55	90	95	94	100	100
"	36	84	93	86	98	96

VISIBLE DETECTION, 627nm

50	19	18	49	36	48	60
"	32	48	62	58	57	68
"	36	56	74	67	68	78
"	27	46	77	67	73	78

205

Recoveries from fortified control milk was determined and results are listed in Table 2. Comparison of the data showed good agreement between the two detectors. While the recovery of thionin was low, it was sufficient for this study. No attempt was made to optimize the recovery of any of the dyes. Methodology was kept simple to minimize oxidation of any leuco-forms of metabolites, if present, by reducing sample manipulations. This data therefore could give a more accurate metabolic mechanism.

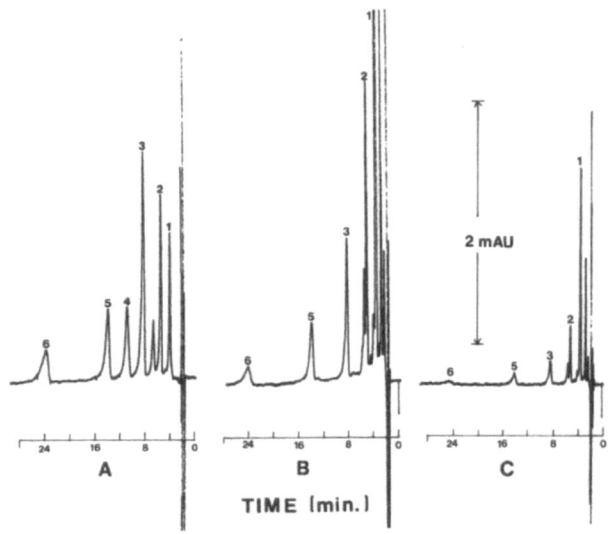

FIGURE 8. · TYPICAL CHROMATOGRAM – VISIBLE DETECTION, 627NM, 0.005 AUFS. A –MBMIX, 0.1ug/mL EACH IN METHANOL, 50uL INJECTION. ORDER OF ELUTION, (1) THIONIN, (2) AZURE C, (3) AZURE A, (4) TOLUIDINE BLUE O, (5) AZURE B AND (6) METHYLENE BLUE. B – 8hr POST-DOSING MB-INCURRED MILK, 10mL THROUGH METHOD, FINAL DILUTION IN 5.0mL IN MOBILE PHASE, 50uL INJECTION. C – 1.0mL ALIQUOT OF B WAS DILUTED TO 2.0mL WITH MOBILE PHASE, 20uL INJECTION.

The procedure was applied to MB-incurred milk samples (8hr and 24hr post-dosing) (Table 3). These samples showed (Munns et al) that MB rapidly depleted in milk to its DMM. Our data agreed very well with his for levels of MB, azure A and azure B these samples. Additionally, we detected azure C and thionin in the 8hr sample at levels >100ppb.

Based on our data for these analytes, both detectors appear to perform satisfactorily. From the stand point of sensitivity, ED has the potential of an increase of ten(10) fold as compared to a five(5) fold increase of visible detection based the current range setting of 10nAFS and 0.005AUFS, respectively. The detectors were set-up in series. The UV/VIS detector was placed upstream (first) and the ED in the downstream (second) position. This scheme produced a slight tailing affect at ED detector due to the extra volume created from the UV/VIS detector cell.

This chromatographic system performed well for both detectors when the k' of the analyte was >4. In the case of thionin and azure C in milk , their k' should be increased through mobile phase modification to remove them from the early eluting elements of the milk matrix. When the target analyte has been determined and a chromatographic system is optimized for that compound, we believe either detection system would perform well. Acceptance of ED has been slow but its application to this and/or any other similar project produces additional and valuable information not available with other techniques.

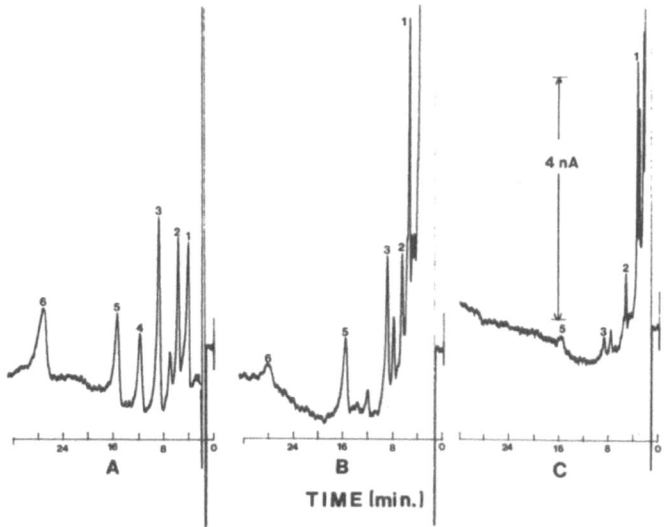

FIGURE 9. TYPICAL CHROMATOGRAM - ELECTROCHEMICAL DETECTION, +1.000V vs
Ag/AgCl, 10nAFS.   A -MBMIX, 0.1ug/mL EACH IN METHANOL, 50uL INJECTION.
ORDER OF ELUTION, (1) THIONIN, (2) AZURE C, (3) AZURE A, (4) TOLUIDINE BLUE
O, (5) AZURE B AND (6) METHYLENE BLUE. B - 8hr POST-DOSING MB-INCURRED
MILK, 10mL THROUGH METHOD, FINAL DILUTION IN 5.0mL IN MOBILE PHASE, 50uL
INJECTION. C - 1.0mL ALIQUOT OF B WAS DILUTED TO 2.0mL WITH MOBILE PHASE,
20uL INJECTION.

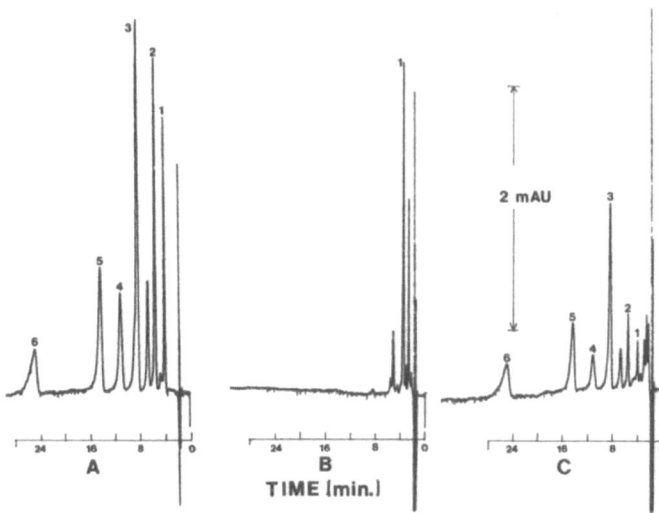

FIGURE 10. TYPICAL CHROMATOGRAM - VISIBLE DETECTION, 627NM, 0.005 AUFS.
A -MBMIX, 0.1ug/mL EACH IN METHANOL, 50uL INJECTION.   ORDER OF ELUTION, (1)
THIONIN, (2) AZURE C, (3) AZURE A, (4) TOLUIDINE BLUE O, (5) AZURE B AND (6)
METHYLENE BLUE. B - 24hr POST-DOSING MB-INCURRED MILK, 10mL THROUGH METHOD
FINAL DILUTION IN 5.0mL IN MOBILE PHASE, 30uL INJECTION. C - CONTROL MILK
FORTIFIED @50PPB OF EACH ANALYTE, 10mL THROUGH METHOD, FINAL DILUTION IN
5.0mL MOBILE PHASE, 50uL INJECTION.

TABLE 3.   **RESULTS FROM INCURRED MILK SAMPLES**

ELECTROCHEMICAL DETECTION, +1.000V vs Ag/AgCl

POST-DOSING	THIO	AZ C	AZ A	TOL.B	AZ B	MB
8hrs	N/DTR	123	---	N/DTC	26	30
	"	135	---	"	28	32
	"	163	---	"	35	40
	430	170	27	---	32	43
	420	180	28	---	35	36
	390	150	25	---	32	36
24hrs	342	94	4	---	0	0
	350	81	6	---	0	0
	328	77	8	---	0	0
	272	66	6	---	0	0

VISIBLE MODE DETECTION, 627nm

POST-DOSING	THIO	AZ C	AZ A	TOL.B	AZ B	MB
8hrs	N/DTR	70	18	N/DTC	29	32
	"	73	19	"	31	35
	"	77	20	"	30	35
	340	98	24	"	35	41
	350	92	23	"	37	41
	300	91	22	"	33	41
24hrs	180	32	0	0	0	0
	168	28	0	0	0	0
	168	30	0	0	0	0
	145	25	0	0	0	0

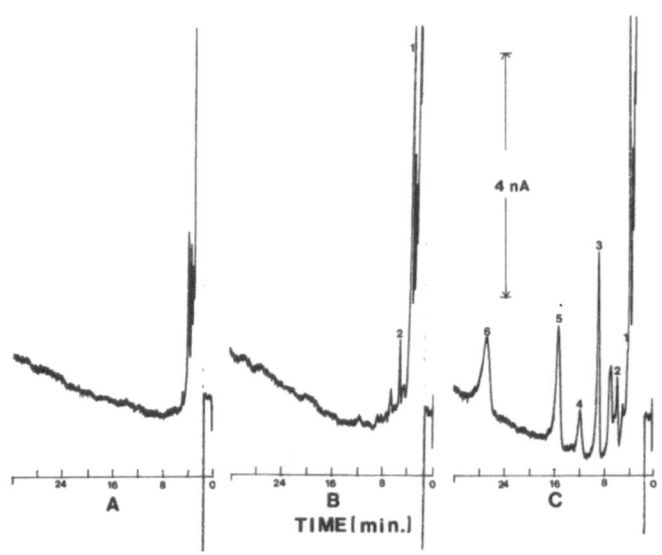

FIGURE 11. TYPICAL CHROMATOGRAM - ED DETECTION, +1.000V vs Ag/AgCl, 10nAFS. A - CONTROL MILK, 10mL CONTROL MILK THROUGH METHOD. FINAL DILUTION, 5.0mL MOBILE PHASE, 20uL INJECTION. B - 24hr POST-DOSING MB-INCURRED MILK, 10mL THROUGH METHOD FINAL DILUTION, 5.0mL IN MOBILE PHASE, 20uL INJECTION. C - CONTROL MILK FORTIFIED @50PPB OF EACH ANALYTE, 10mL THROUGH METHOD, FINAL DILUTION, 5.0mL MOBILE PHASE, 50uL INJECTION. ELUTION, (1) THIONIN, (2) AZURE C, (3) AZURE A, (4) TOLUIDINE BLUE O, (5) AZURE B AND (6) METHYLENE BLUE.

ACKNOWLEDGEMENTS

The authors wish to thank Dr. David D. Wagner, Chief, Animal Nutrition Branch, Division of Veterinary Medical Research, Center for Veterinary Medicine, FDA, Beltsville, MD, for preparing and furnishing all incurred milk necessary for this work.

REFERENCES

1. G. Balansard, G. Schwadrohn, E. Vidal-Ollivier and R. Elias, "Thin-layer and High-performance liquid chromatography of tetramethylthionine chloride (methylene blue), used in general medicine and of tetramethylthionine base for ophthalmology", Ann. Pharm. Fr., 46(2):129-132(1988)

2. M. Kanesato, K. Nakamura, O. Nakata and Y. Morikawa, "Analysis of ionic surfactants by HPLC with ion-pair extraction detector", J. Am. Oil Chem. Soc., 64(3):434-438(1987)

3. L. Gagliardi, G. Cavazzutti, A. Amato, A. Basili and D. Tonelli, "Identification of cosmetic dyes by ion-pair reverse-phase high-performance liquid chromatography", J. Chromatogr., 394(2):345-352(1987)

4. A. R. DiSanto and J. G. Wagner, "Pharmacokinetics of highly ionized drugs I: Methylene blue - whole blood, urine and tissue assays", J. Pharm. Sci., 61(4):598-602(1972)

5. A. R. DiSanto and J. G. Wagner, "Pharmacokinetics of highly ionized drugs II: Methylene blue - Absorption, metabolism and excretion in man and dog after oral administration", J. Pharm. Sci., 61(7):1086-1090(1972)

6. J. Watanabe and R. Fujita, "Elimination of methylene blue in dogs after oral or intravenous administration", Chem. Pharm. Bull., 25(10):2561-2567(1977)

7. J. Watanabe and K. Mori, "Small intestinal absorption of methylene blue in rats, guinea pigs and rabbits", Chem. Pharm. Bull., 25(6):1194-1201(1977)

8. G. Ziv and J. E. Heavner, "Permeability of the blood-milk barrier to methylene blue in cows and goats", J. Vet. Pharmacol. Ther.,7(1):55-59(1984)

9. A. Mizuo and K. Hiraga, "Mutagenicity of dyes in the microbial system", Tokyo Toritsu Eisei Kenkyusho Kenkyu Nempo,27(2):153-158(1976)

10. A. S. N. Murthy and A. P. Bhardwaj, "Charge-transfer interaction of thionine and Toluidine blue with amines", J. Chem. Soc. Faraday Trans. 2,78(11):1811-1814(1982)

11. P. N. Marshall, "The composition of stains produced by the oxidation of Methylene Blue", Histochem. J.,8(4):431-442(1976)

12. M. R. McKamey and L. A. Spitznagle, "Chromatographic mass spectral and visible light absorption characteristics of toluidine blue O and related dyes", J. Pharm. Sci.,64(9):1456-1462(1975)

13. R. Bonneau, J. Joussot-Dubien and J. Faure, "Mechanism of photoreduction of thiazine dyes by EDTA studied by flash photolysis I", Photochem. Photobiol.,17(5):313-319(1973)

14. F. Sierra, C. Sanchez-Pedreno, T. Perez Ruiz, M. Hernandez Cordoba and Martinez Lozano, C., "Oxidimetric determination of photochemically reduced triazine dyes. Amperometric evaluations"An. Quim., 72(5):456-459(1976)

15. J. E. Roybal, R. K. Munns, J. A. Hurlbut and W. Shimoda, "High-performance liquid chromatography of Gentian violet, its demethylated metabolites, leucogentian violet and methylene blue with electrochemical detection", J. Chrom.,467:259-266(1989)

16. J. E. Roybal, R. K. Munns, J. A. Hurlbut and W. Shimoda, "Determination of Gentian violet, its demethylated metabolites and leucogentian violet in chicken tissue by liquid chromatography with electrochemical detection", J. Assoc. Off. Anal. Chem., 73(6):940-946(1990)

17. R. K. Munns, J. E. Roybal, J. A. Hurlbut and W. Shimoda, "Rapid method for the determination of leucogentian violet in chicken fat by liquid chromatography with electrochemical detection", 73(5):705-708(1990)

18. R. K. Munns, D. C. Holland, J. E. Roybal, J. G. Meyer, J. A. Hurlbut and Long, A. R., " Liquid chromatographic determination of methylene blue and its metabolites in milk" submitted for publication

19. W. W. Dean, G. J. Lubrano, H. G. Heinsohn and M. Stastny, "The Analysis of Romanowsky Blood Stains by High Performance Liquid Chromatography", J. of Chrom., 124:287-301(1976)

THE PRESENCE OF DIHYDROERYTHROIDINES IN THE MILK OF GOATS FED
ERYTHRINA POEPPIGIANA AND *E. BERTEROANA* FOLIAGE

Lori D. Payne and Joe P. Foley

Department of Chemistry
LSU, Baton Rouge, LA. 70803

INTRODUCTION

The use of nitrogen fixing crop species, such as alfalfa and clover, as a high protein animal feed is an established practice in U.S. agriculture. However, in many tropical countries nitrogen fixing pasture species are not commonly used. Temperate legumes do not grow well due to a combination of climate, competition and lack of traditional commercial inputs. On the other hand, nitrogen fixing tree species are abundant in the tropics. Many have been introduced or naturalized as a result of the coffee and cocoa industry where crops are managed below a prunable canopy of trees. Innovative uses of these trees have been developed by several international development organizations such as the Centro Agronómico Tropical de Investigación y Enseñanza (CATIE) in Costa Rica.

One promising nitrogen fixing tree genus, *Erythrina*, has been used successfully in several agroforestry systems, for example as live fences, as shade for coffee plantations, as guides for climbing crops and more recently as a feed supplement for animals. Several studies by the Nitrogen Fixing Tree Projoect at CATIE have indicated that the foliage *Erythrina poeppigiana* and *E. berteroana* can substitute for all or a portion of the protein required by a ruminant with no detectable adverse effects (1, 2, 3). On the other hand, *Erythrina* is not a suitable feed for monogastric animals. Several non-ruminant species demonstrated toxic symptoms in feeding studies with *Erythrina*. Rats fed a diet consisting of 51% (w/w) *E. berteroana* lost weight over a 28 day experiment while the controls more than tripled their weight (4) and pigs fed a diet containing less than 3% *E. poeppigiana* failed to gain weight and where removed from a study due to their poor physical condition (2). Sterility and death resulted in a study involving rabbits fed *Erythrina* (5). Additional evidence that *Erythrina* may be toxic to non-ruminants animals lies in its use as a medicinal herb used to combat insomnia in humans and its use by South American Indians as a fish poison (6).

Analysis of Antibiotic Drug Residues in Food Products of Animal Origin
Edited by V.K. Agarwal, Plenum Press, New York, 1992

One explanation of the toxicity differences between ruminant and monogastric animals with respect to *Erythrina* species is that some toxic compound(s) in the foliage of the *Erythrina* species may be detoxified by the microorganisms in the rumen and may not detoxified by the gastrointestinal tract of the monogastric animals. Alternatively, some compound(s) may be metabolically activated in the monogastric animals and not in the rumen of the ruminants. The reductive neutral atmosphere of the rumen may generate very different metabolic reactions than the acidic stomach of the monogastric animal.

The genus *Erythrina* is known to contain curare like alkaloids (6). Clinically active alkaloids were isolated from the seeds of various *Erythrina* species in the 1930s and 1940s by Folkers and coworkers at Merck Pharmaceutical Company (7, 8). These alkaloids were of interest because they retained neuromuscular transmission blocking activity even when administered orally. Further studies resulted in the isolation of several other "*Erythrina*-like" alkaloids from the seeds of various *Erythrina* species. The majority of new *Erythrina* alkaloids were less or as potent as the original drug, ß-erythroidine (Figure 1). In the 1980s, ß-erythroidine was isolated from the leaves of *E. poeppigiana* and *E. berteroana* (9). Previous work in our lab confirmed that ß-erythroidine was not only present in *E. berteroana* and *E. poeppigiana*, but was the major alkaloid present in these species (10).

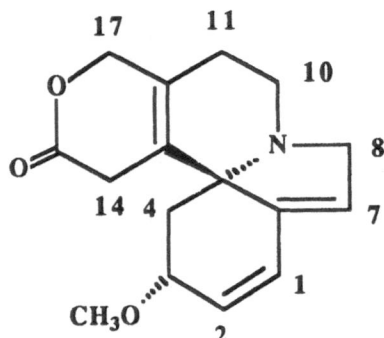

Fig. 1. ß-Erythroidine

There have been reports of alkaloids detected in the milk of animals fed alkaloid containing plants. In the pioneering days of the United States, a disease known as milk sickness caused many fatalities. The disease was traced to the consumption of milk from cows that had eaten white snakeroot (*Eupatorium rugasum*) (11). In a more recent case, anagyrine present in the milk of diary animals which consumed *Lupinus latifolius*, may have caused birth defects in California (12).

Much of the recent work in milk transfer of toxins has centered on the species Tanzy Ragwort (*Senecio jacobaea*), a poisonous plant found in the Pacific Northwest known to contain hepatoxic pyrrolizidine alkaloids. In one study (13), histopathologic changes were not detected in the calves consuming milk of the cows eating *S. jacobaea*, but clinical chemical tests showed hepatic biochemical lesions in calves. Rats fed goat's milk which contained 7.5 ng of pyrrolizidine alkaloids developed swollen hepatocytes of centrilobular distribution and biliary hyperplasia (14). In another study in rats there was a high rate of transfer of the toxic substances through the milk as measured by lesions in the young. The dose given to the mother was not lethal but was lethal to the suckling young (15). Additional studies also indicate that *Senecio* alkaloids can reach the milk of animals that consume *Senecio* spp. (16). There is evidence that other types of alkaloids, namely ergot, indolizidine and quinolizidine alkaloids also may contaminate the milk of dairy animals (11).

Milk transfer of toxins can occur with mycotoxins and ketones as well as the alkaloids mentioned (11). In many instances, these natural toxins in the milk of dairy animals pose no threat to the general public in countries with a developed dairy industry because the toxins are highly diluted with uncontaminated milk at processing centers before distribution. However, toxins are of concern to the small family farmer in developing countries who may use only one or two animals to provide milk for the entire family over a long period of time. In this case, any toxins consumed by the dairy animals may concentrate in the milk and pose a health threat to the family. Since *Erythrina* foliage is known to contain toxic alkaloids which cause muscle paralysis, it is imperative that the milk of the animals consuming *Erythrina* foliage be examined for alkaloid content and for the presence of toxic metabolites.

EXPERIMENTAL

Feeding Study

Twenty dairy goats were preselected from the herd at the Animal Production Department, Centro Agronómico Tropical de Investigación y Enseñanza (CATIE) in Costa Rica according to due date. The goats were crosses of Nubian, Tottenberg, Saanen and Alpine and ranged in age from two to eight years. One month before expected parturition, the selected animals were fed a diet of 1 kg concentrate with grass and banana ad lib. This diet served as the control diet during the study. After parturition, the animals were separated into treatment groups. Treatment I was the control diet. Treatment II consisted of a diet of 40% by weight *E. poeppigiana* foliage and 60% grass ad lib. with no concentrate. Treatment III consisted of a diet of 40% by weight *E. berteroana* foliage and 60% grass ad lib. The animals were conditioned on these diets for one month before milk samples were taken.

Data collection related to the feeding experiment occurred between October, 1989 and January 1990; feed consumption was recorded daily, body weights were recorded monthly, milk production was recorded weekly and milk samples were taken weekly for four weeks and then monthly for two months. For the milk analysis, a representative sample of 100 mL of milk was taken from the total daily milk production of each goat, immediately frozen and then lyophilized for storage until analyzed.

Milk Analysis

The lyophilized sample representing 100 mL of milk was placed in a Soxhlet apparatus and extracted with 400 mL of hexane for 24 hours. The solvent was evaporated and the residue dissolved in 20 mL of 1% H_2SO_4. The acid solution was basified to pH 10 with $NaHCO_3$ and 25% NH_4OH and extracted with 3x10 mL chloroform and dried over Na_2SO_4. After solvent evaporation, the residue was dissolved in enough chloroform to provide a 2-10 mg/mL solution for GC/MS analysis.

The material remaining in the Soxhlet device after the hexane extraction was extracted with 400 mL ethanol. After rotoevaporation, the residue was dissolved in 50 mL H_2SO_4 and extracted with 4x10 mL chloroform which was discarded. The acidic solution was basified to pH 10 with $NaHCO_3$ and 25% NH_4OH and extracted with 3x10 mL chloroform and dried over Na_2SO_4. After solvent evaporation, the residue was dissolved in enough chloroform to provide a 2-10 mg/mL solution for GC/MS analysis.

The basic solution remaining after chloroform extraction was re-acidified to pH<2 with 25% HCl, heated to 60°C over a water bath for 1 hour and left overnight to encourage hydrolysis. This fraction was worked up for GC/MS analysis as described in the preceding paragraph.

Rumen Incubations

Dried plant material (0.5 g) was incubated with 50 mL of a 20% rumen liquor/80% artificial saliva solution in sealed tubes flushed with CO_2 in an incubator at 40°C and under an atmosphere of CO_2 for 48 hours. The rumen liquor was obtained by squeezing the rumen contents of a fistulated goat through a cheese cloth. The artificial saliva was made from 9.8 g $NaHCO_3$, 7.0 g Na_3PO_4, 5.7 g KCl, 4.7 g NaCl and 1.2 g $MgSO_4$ in 10 L of distilled water. 1 mL of 5.3 g $CaCl_2$ in 100 mL of water was added for each liter of mixed rumen liquor solution. The incubations were terminated by the addition of 5 drops of 25% HCl. The contents of the tubes were rinsed into lyophilization flasks, immediately frozen and then lyophilized for further analysis. The lyophilized rumen samples were extracted and analyzed according to the procedure described in the milk analysis section.

GC/MS Analysis

1 μL of the extract dissolved in chloroform was run on a Hewlett Packard 5890 gas chromatograph and analyzed with an Hewlett Packard 5971 quadrupole mass spectrometer. Gas chromatographic conditions were as follows: 40°C for 3 min., ramp of 20°C/min. to 250°C and held at 250°C for 25 minutes. The column was a DB-5 (J&W Scientific) 30 m capillary column

with a film thickness of 0.25 μm and an i.d. of 0.25 mm. The
flow rate was 30 cm/sec. of helium. The data was acquired and
analyzed using DOS based ChemStation (Hewlett Packard)
software.

Synthetic Hydrogenation

15 to 30 mg of ß-erythroidine was dissolved in 1 mL of
water to which 62 mg of solid NaOH and 1 g of raney nickel
(Aldrich) was added. The solutions were either left overnight
stirring or placed under 3 atm. of H_2 for 90 min. The mixture
was filtered through 0.45 μm nylon filter, rinsed, acidified
and left overnight. The acidic solution was basified to pH 10
with 25% NH_4OH and extracted with 5x5 mL chloroform, which was
then evaporated and the resulting residue dissolved in enough
chloroform to make a solution of 2-10 mg/mL for GC/MS
analysis.

Statistical Analysis

Statistical analysis was performed using Statview 512+
(Abacus Concepts) on a Macintosh computer.

RESULTS

Through the use of mass spectral interpretation and
comparison to the standard ß-erythroidine, dihydroerythroidine
isomers and tetrahydroerythroidine were identified. The
dihydro- and tetrahydroerythroidines were detected in the milk
of the experimental animals in the two treatment groups
consuming *E. poeppigiana* and *E. berteroana* foliage and in the
rumen incubation solutions. The tabulated mass spectra of
these compounds and that of ß-erythroidine are presented in
Table 1. The various reduced compounds were well resolved on
the DB-5 column and represented a very small portion of the
total residue from the milk samples. The majority of the
compounds in this fraction were long chain aliphatic compounds
which were also well resolved. Three dihydroerythroidines
(DHEI, DHEII and DHEIII) and one tetrahydroerythroidine (THE)
were detected in the milk and rumen samples. DHEI constituted
the major portion of the DHEs in the milk and rumen samples,
DHE II was present in smaller amounts and DHEIII was present
in trace quantities, if at all. Some unreduced ß-erythroidine
was detected in two of the milk samples and in two of the
rumen samples. No erythroidine or dihydroerythroidine
compounds were detected in the milk collected from the goats
in the control group on a diet free of *Erythrina* foliage.

Two dihydroerythroidine (DHEIV and DHEV) isomers were the
only products of the synthetic hydrogenations over raney
nickel. DHEIV was the major isomer. No parent ß-erythroidine
or tetrahydro- compounds were detected. The mass spectra of
major DHE isomers from the biological hydrogenation (DHEI) and
the major isomer from the synthetic hydrogenation (DHEIV) are
presented in Figure 2. The synthetic dihydroerythroidines
were used to estimate the amount of dihydroerythroidines in
the milk samples using the areas in the total ion chromatogram
and background subtraction. Figure 3 displays the results of
the milk analysis over time on a treatment averaged basis.

Table 1. Tabulated mass spectral data as percent of base peak of biologically hydrogenated dihydroerythroidines (DHEI, DHEII and DHEIII),catalytically hydrogenated dihdyroerythroidines (DHEIV and DHEV), tetrahydroerythroidine (THE) and ß-erythroidine (E).

m/z	DHEI	DHEII	DHEIII	DHEIV	DHEV
			-----Percent of Base Peak-----		
275	20	42	46	16	37
260	58	91	94	1	10
245	38	31	20	7	16
244	44	76	66	23	81
230	100	93	95	3	2
217	5	49	59	100	49
216	19	74	74	8	21
204	17	20	6	2	2
200	15	20	24	7	4
191	6	62	31	<1	1
186	21	38	25	2	4
178	13	24	50	<1	2
172	19	100	100	24	100
158	16	56	50	5	34
146	15	20	17	2	7
132	15	54	39	9	40
130	14	33	20	9	9
77	28	45	50	4	6

THE	m/z	%	E	m/z	%
	277	9		273	55
	262	18		258	26
	247	14		242	100
	232	17		240	27
	218	6		214	14
	206	8		198	17
	178	29		184	14
	164	10		182	21
	148	100		170	21
	146	16		156	15
	132	18		130	34
	130	5		91	20
	77	14		77	25

Statistical analysis (t-tests) revealed no difference between the two *Erythrina* treatment groups with respect to the concentration of DHEs even though the concentration of DHEs in the *E. poeppigiana* treatment group was approximately twice that of the *E. berteroana* group. The *Erythrina* treatment groups were significantly different from the control groups in the concentration of DHE in most instances, but not in all due to the variability of the data. There was no correlation between DHE concentration in the milk and fat content of the milk (correlation coefficient = -0.192) or the amount of milk produced (correlation coefficient = -0.186). There was no

significant difference at the 95% confidence level (c.l.) with
respect to milk production between the treatment groups and
the control group until the later part of the study (Figure
4).

There was a significant difference (95% c.l.) between all
groups with respect to the amount of feed consumed. The diet
containing *E. poeppigiana* was consumed the in the greatest
quantity, followed closely by the *E. berteroana* group. The
control group ate significantly less foliage than either
treatment group. The contribution of the concentrate was not
included in the calculations of the total amount of foliage
consumed for the control group.

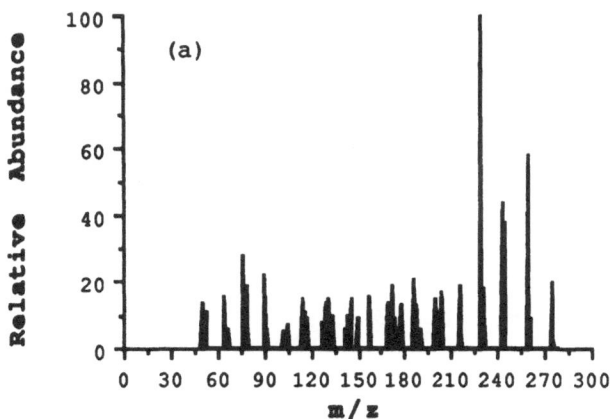

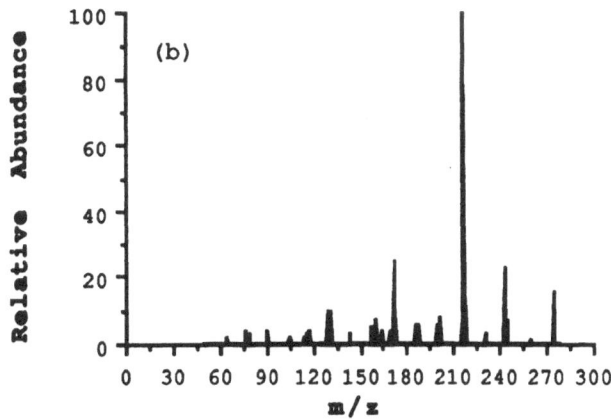

Fig. 2. Reconstructed mass spectra of the primary
         biological dihydroerythroidine isomer, DHEI
         (a), and the primary synthetic dihydro-
         erythroidine isomer, DHEIV (b).

DISCUSSION

The mass spectra of the dihydroerythroidines and
tetrahydroerythroidine detected in the rumen samples match
exactly the dihydro- and tetrahydroerythroidines detected in
the milk samples of the treatment groups of each species.
These compounds were not detected in the milk of the control
group. The ß-erythroidine present in the foliage of the *E.
poeppigiana* and *E. berteroana* is evidently hydrogenated in the
rumen of the dairy goats and transported to the milk of these
animals.

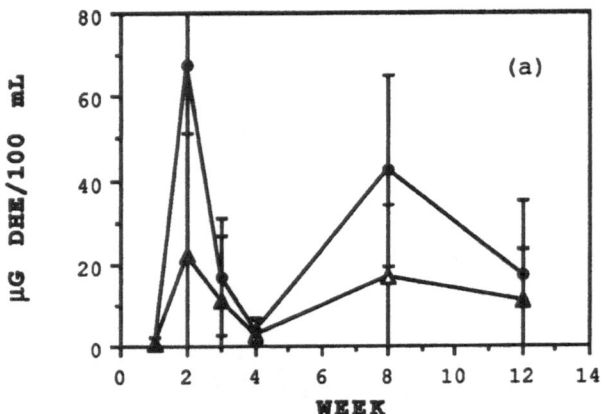

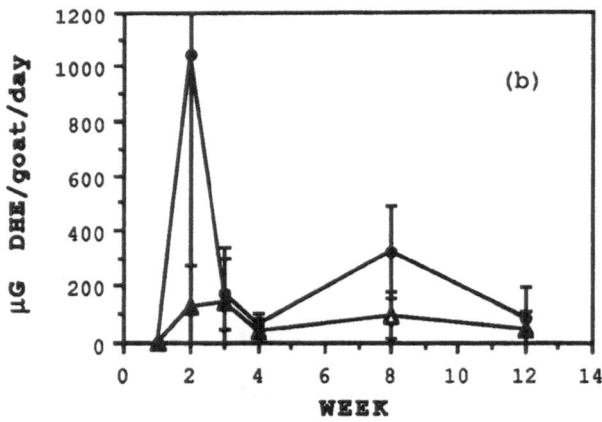

Fig. 3. Amount of dihydroerythroidines (DHEs) present in the
milk of goats which consumed *E. poeppigiana* (•) and
E. berteroana (Δ) and foliage by goat species. (a)
µg per 100 mL of milk; (b) total µg DHE produced per
goat per day.

In the rumen environment at a pH of 5.5 to 7.0, the absence of oxygen and the presence of microorganisms promotes reductive reactions (17). Hydrogenation is one of many reductive chemical reactions that takes place in the rumen. Most reports of hydrogenation focus on the rapid hydrogenation of unsaturated fatty acids by rumen bacteria for the construction of bacterial cells (18). During this type of hydrogenation, double bond migrations are known to occur (19).

The mass spectral data of the biologically formed DHEs and the synthetic DHEs indicate that the two hydrogenation processes produce different DHE isomers. The biologically hydrogenated DHEs have major molecular ion peaks as well as major peaks corresponding to $(M-CH_3)^+$ and $(M-OCH_3)^+$. The base peak of DHEI (m/z = 230) corresponds to a loss of an additional methylene group as well as a loss of OCH_3. These results are consistent with the fragmentation pattern of dienoid type of *Erythrina* alkaloids proposed by Boar and Widdowson (20). If the conjugated diene is intact after hydrogenation, then the double bond present in the lactone ring of ß-erythroidine (Figure 1) is hydrogenated in the rumen which may result in the formation of two diastereomers.

Inspection of the mass spectra of the major synthetic isomer, DHEIV, reveals that the molecular ion and $(M-CH_3)^+$ peaks are weak when compared to the mass spectra of the biological DHEs and that the base peak of m/z = 217 represents a loss of 58 mass units. This loss is typical of a alkenoid type *Erythrina* alkaloids (20). Hydrogenation of the double bond at the 2-3 position, probably the most facile position to synthetically hydrogenate, and subsequent double bond migration would result in a double bond at the 1-6 position.

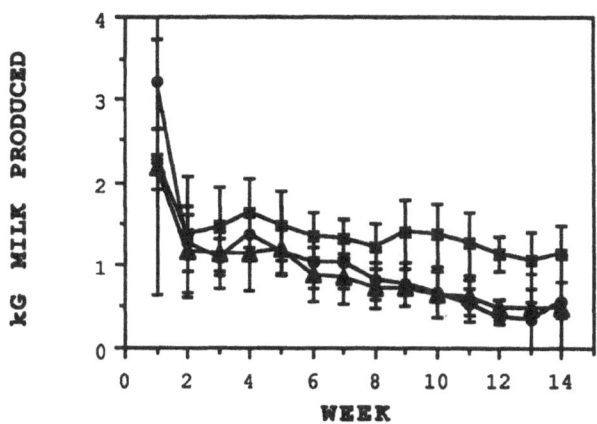

Fig. 4. The amount of milk produced (kg) by each treatment group. (×) control; (•) *E. poeppigiana*; (Δ) *E. berteroana*.

A retro-Diels Alder reaction would result in a loss of 58 units and the second mechanism of fragmentation proposed by Boar and Widdowson would also result in a loss of 58 units. These results support the hypothesis that although catalytic and bacterial hydrogenations are similar in many instances, conjugated double bonds are resistant to microbial hydrogenation (21). It should be noted that the pharmacological studies done by Unna (8) and others utilized DHE formed upon catalytic hydrogenation of ß-erythroidine. Therefore it is likely that the synthetic dihydroerythroidine investigated in the pharmacological and toxicological drug studies of the 1940s also contained a double bond at the 1-6 position and the 12-13 position.

The wide variability in the concentration of dihydroerythroidines in the milk of the treated animals was expected. Neither the experimental animals nor the foliage used as feed was genetically homogeneous in the feeding study. This variability may be environmental, genetic or biochemical in nature. The alkaloid content of the leaves may vary with environmental conditions thus causing variability in the amount of ß-erythroidine the animals were consuming. It is interesting that both *Erythrina* species varied similarly. The bacteria in the rumen of each animal may be slightly different resulting in different rates of hydrogenation of the ß-erythroidine and/or competing biochemical reactions in the rumen. Alternatively, each animal may metabolize the dihydroerythroidine produced differently or excrete them at different rates. Other contributions to the variability can be hypothesized. These results emphasize the importance of utilizing genetically homogeneous material in scientific investigations, particularly in feeding studies where biological variation may be great.

Only a small fraction of the ß-erythroidine consumed may reach the milk in an unaltered or hydrogenated form. Studies done with *Erythrina* clones have shown that *Erythrina* foliage is approximately 0.1% by weight of ß-erythroidine (10). If the goats eat on the average of 2.2 kg of *Erythrina* foliage per day, then approximately 440 mg of ß-erythroidine are consumed per day per goat assuming a dry weight of 20%. Estimates of the total amount of dihydroerythroidines in the milk are between 1 µg and 1 mg per day per goat. So the DHE found in the milk represents less than 0.2% of the total amount consumed if the above assumptions are used. It is possible that the foliage used in the feeding study was exceptionally low in alkaloid content, however there may be other explanations for the possibly low amount of DHEs detected in the milk of the experimental animals, such as environmental variation and differences in bacterial composition in the rumen of the experimental animals as previously mentioned.

The results of the rumen incubations suggest that almost all the ß-erythroidine found in the foliage is metabolized. Very little parent compound is detected in the rumen solutions after 48 hours, so it is unlikely that ß-erythroidine is eliminated through the feces. Since the dihydro- compounds are probably transported by the circulatory system at a pH at which they are soluble and because they have no particular affinity for lipids, it is possible that the DHE are eliminated through the urine as well as the milk. The parent compound, ß-erythroidine, is known to be soluble in blood serum as intravenous injection as a method of drug introduction in clinical use of the drug demonstrates. There have been no studies done on the degradative metabolism of ß-erythroidine in humans or animals, however clinical use of the drug suggests that ß-erythroidine is rapidly degraded or eliminated in human subjects (22). There was no correlation between the milk fat content and the amount of DHE detected suggesting that the DHEs are not preferentially sequestered in the fat of the animal. No *Erythrina* alkaloids were detected in the hexane fraction during extraction. In addition, most of the treatment animals maintained or lost weight, so fat deposition is an unlikely method of elimination.

The ß-erythroidine may be degraded by rumen bacteria into compounds not present in the alkaloid fraction analyzed by GC/MS. Although several types of heterocyclic alkaloids, for example pyrrolizidine and indole alkaloids (23), are not substantially degraded by ring fission in the rumen, there are examples of heterocyclic fissions of other types of compounds such as coumarins where the lactone moiety is attacked (24) or diquat which is converted to a methyl urea derivative and CO_2 (25). So it is possible that much of the drug is degraded and excreted before it reaches the milk.

It is difficult to assess the human health risk associated with the discoveries presented in this paper. Dihydroerythroidine is one of the most potent forms of ß-erythroidine with an effective dose for complete paralysis in mice of 80 µg/kg (26). This amount was exceeded in many milk samples on a total amount of DHEs produced per goat per day basis (Figure 3). However, mass spectral data presented here suggest that the DHEs formed in the rumen are different from the synthetic DHE tested in toxicological studies in the 1940s and 1950s (27) and therefore, the pharmacological data currently available in the scientific literature on the toxicity of DHE may not be pertinent to the isomers detected in the milk and rumen of the dairy goats in this study. From a review of the synthetic studies done on synthetically hydrogenated erythroidine derivatives, Sniecknes concluded that in the case of ß-erythroidine, the pharmacological properties of the drug were not dependent on the double bond configuration of the diene in the C and D rings (28). Several different optical isomers showed similar activities in biological assays. Furthermore, long term chronic exposure to these alkaloids, as well as other types alkaloids (29), which may occur when a family consumes the contaminated milk regularly, is unknown.

In conclusion, ß-erythroidine, a paralyzing drug which is present in the foliage of *Erythrina berteroana* and *E. poeppigiana* tree species, is rapidly hydrogenated in the rumen of dairy goats. Erythroidine, dihydroerythroidines and tetrahydroerythroidine were detected in the milk of the two treatment groups fed *E. poeppigiana* and *E. berteroana* foliage. The DHEs detected in the milk were not the same isomers formed upon catalytic hydrogenation of ß-erythroidine and therefore the human health hazard associated with the consumption of milk which contains biologically hydrogenated DHEs is uncertain.

ACKNOWLEDGMENTS

The authors wish to thank CATIE for their cooperation during the course of this project and for a travel award for LDP. The gift of ß-erythroidine from Dr. V. Boekelheide of the University of Oregon is gratefully acknowledged.

REFERENCES

1. C. Samur Rivero, "Producción de Leche de Cabras Alimentadas con King Grass (*Pennisetum purpureum*) y Poró (*Erythrina poeppigiana*), suplementads con Fruto de Banano (*Musa* sp. cv. 'Cavendish)," Master's Thesis, Centro Agronónmico Tropical de Investigación y Enseñanza, Turrialba, Costa Rica (1984).
2. A. Vargas Fournier, "Estudio Preliminar: Uso de la Harina de Pescado y Hojas de Poró (*Erythrina poeppigiana*) en el Desarrollo y Engorde de Cerdos Alimentados con Banano," Centro Agronónmico Tropical de Investigación y Enseñanza, Turrialba, Costa Rica (1983).
3. J. E. Benavides, Utilizacion de Follaje de Poró (*Erythrina poeppigiana*) para Alimentar Cabras Bajo Condiciones de Tropico Húmedo," Segundo Congreso de las Asociación Mexicana de Zootecnistas y Técnicos en Caprinocultura, Mazatlán, México (1986)
4. R. M. De Leon I. de Pineda, "Evaluacióna Nutricional de la Harina del Arbol de Pito (*Erythrina berteroana* Urban, Symb), en Animales de Laboratorio," Masters Thesis, Universidad de San Carlos de Guatemala, San Carlos, Guatemala (1984).
5. G. W. Martin, Edible leaves from nitrogen fixing trees, <u>Nitrogen Fixing Trees Research Reports</u>, 2:57 (1984).
6. S. F. Dyke and S. N. Quessy, "*Erythrina* and related alkaloids, <u>in</u>: "The Alkaloids," R. H. F. Manske, ed., Academic Press, New York, 18:1 (1981).
7. K. Folkers and R. T. Major, Isolation of erythroidine, an alkaloid of curare action, from *Erythrina americana* Mill., <u>J. Am. Chem. Soc.</u>, 59:1590 (1937).
8. K. Unna, M. Kniazuk and J. G. Greslin, Pharmacologic action of *Erythrina* alkaloids I. ß erythroidine and substances derived from it, <u>J. Pharmacol. Exper. Therap.</u>, 80:39 (1944).
9. A. H. Jackson and A. S. Chawla, Studies of *Erythrina* aklaloids, Part IV. G.C./M.S. investigations of aklaloids in the leaves of *E. poeppigiana, E. macrophylla, E. berteroana,* and *E. saliviiflora,* <u>Allertonia</u>, 3:39 (1982).
10. L. D. Payne and J. P. Foley, Gas Chromatography and mass spectrometry of *Erythrina* alkaloids from the foliage of genetic clones of three *Erythrina* species, <u>in</u>: Chromatography and Pharmaceutical Analysis, S. Ahuja, ed., American Chemical Society Symposium Series (1992). (Submitted)
11. P. R. Cheeke and L. R. Shull, "Natural Toxicants and Poisonous Plants," Avi Publishing Co., Inc., Westport, Connecticut (1985).
12. W. W. Kilgore, D. G. Crosby, A. L. Craigmill and N. K. Poppen, Toxic plants as possible human teratogens, <u>California Agriculture</u>, 35:6. (1981).

13. A. E. Johnson, Changes in calves and rats consuming milk from cows fed chronic lethal doses of *Senecio jacobaea* (Tanzy Ragwort), A. J. Vet. Res., 37:107 (1976).

14. D. E. Goeger, P. R. Cheeke, J. A. Schmitz and D. R. Buhler, Effect of feeding milk from goats fed tansy ragwort (*Senecio jacobaea*) to rats and calves. Am. J. Vet. Res., 43:1631 (1982).

15. R. Schoental, Liver lesions in young rats suckled by mothers treated with the pyrrolizidine alkaloids, lasiocarpine and retrorsine, J. Pathol. Bact., 77:485 (1959).

16. J. O. Dickinson, M. P. Cooke, R. R. King and P. A. Mohamed, Milk transfer of pyrrolizidine alkaloids in cattle, JAVMA, 169:1192 (1976).

17. R. E. Hungate, "The Rumen and Its Microbs," Academic Press, New York (1966).

18. P. F. V. Ward, T. W. Scott and R. M. C. Dawson, The hydrogenation of unsaturated fatty acids in the ovine digestive tract, Biochem J., 92:60 (1964)

19. P. F. Wilde and R. M. C. Dawson, The biohydrogenation of linolenic acid and oleic acid by rumen micro-organisms, Biochem. J., 98:469 (1966).

20. R. B. Boar and D. A. Widdowson, Mass spectra of the *Erythrina* alkaloids: a novel fragmentation of the spiran system, J. Chem Soc. (B), 1970:1591.

21. G. A. Garton, Aspects of lipid metabolism in ruminants, in: "Metabolism and Physiological Significance of Lipids," R. M. C. Dawson and D. N. Rhodes, eds., Wiley, New York (1964).

22. K. Unna, Curare-like action of *Erythrina* alkaloids, Proceedings of the American Physiological Society, 126:P644-P645 (1939).

23. J. R. Carlson and R. G. Breeze, Ruminal metabolism of plant toxins with emphasis on indolic compounds, J. Animal Sci., 58:1040 (1984).

24. W. H. Gutenmann, J. W. Serum and D. J. Lisk, Feeding studies with VCS-438 herbicide in the dairy cow, J. Agr. Food Chem., 20:991 (1972).

25. R. T. Williams, Toxicolgic implications of biotransformation by intestinal microflora, Tox. Appl. Pharmacol., 23:769 (1972).

26. B. H. Robbins and J. S. Lundy, Curare and curare-like c compounds: a review, Anesthesiology, 8:348 (1947).

27. V. Boekelheide, *Erythrina* alkaloids, Record of Chemical Progess, 16:226 (1955).

28. V. A. Snieckus, *Erythrina* and related alkaloids, in: M. F. Grundon, ed., "The Alkaloids," Burlington House, London, 3:180 (1973).

29. J. E. Peterson, The toxicity of *Echium plantagineum*, in: "Plant Toxicology," A. A. Seawright et al., ed., Animal Research Institute, Yeerongpilly, Australia (1985).

ANALYSIS OF SELECTED CHEMOTHERAPEUTICS

AND ANTIPARASITICS

Rainer Malisch

Chemische Landesuntersuchungsanstalt
(State Institute for Chemical Analysis of Food)
Bissierstr. 5
D-7800 Freiburg
Germany

1. INTRODUCTION

1.1 DRUGS AND TOLERANCES

Modern intensive animal breeding demands routine suppression of different diseases, caused e.g. by vira, bacteria, protozoa or fungi. In Germany, about 250 different substances are admitted in roughly 3,000 drugs for the treatment of various diseases in animals used for the production of food (1 - 3). They are different in their chemical structure as well as in their therapeutic effect.

The official food surveillance should be able to determine possible residues of a great deal of the available drugs. As a prerequisite for an appropriate analysis, acceptable residue limits should be fixed to avoid expensive determinations in a range which is not of toxicological concern. These tolerances have been under discussion for years. For the harmonization of the Common Market, the European Community will set tolerances for all drugs used in veterinary medicine by 1997. The proposed maximum residue limits for substances of interest in this paper are:
- sulfonamides 100 µg/kg (parent drug) in meat, all tissues and milk
- chloramphenicol 10 µg/kg in meat, all tissues
- nitrofuranes 5 µg/kg for all residues with intact 5-nitro-structure in meat, all tissues, eggs
- dimetridazole 10 µg/kg for all residues with intact nitro-imidazole structure in meat, all tissues
- ronidazole 2 µg/kg for all residues with intact nitro-imidazole-structure in meat, all tissues
- febantel, fenbendazole, oxfendazole: 1000 µg/kg for the sum of oxfendazole + oxfendazole-sulphone + fenbendazole in liver; 10 µg/kg for this sum in muscle, kidney, fat and milk.

Analysis of Antibiotic Drug Residues in Food Products of Animal Origin
Edited by V.K. Agarwal, Plenum Press, New York, 1992

225

The Joint FAO/WHO Expert Committee on Food Additives (JECFA) evaluated certain veterinary drug residues in food. The Codex Alimentarius Commission is reviewing the proposed Maximum Residue Limits at different steps of its acceptance procedure. These limits differ to some degree from the proposed EC tolerances. This makes it difficult for an analyst to develop internationally appropriate methods: Generally it is more difficult to establish methods with lower limits of determinations.

1.2 MULTI-METHODS

To avoid an inefficient analysis by applying methods detecting only single substances, multi-residue methods are very useful for the analysis. Comprehensive multi-methods can detect different groups of substances simultaneously, e.g. sulfonamides, chloramphenicol and some antiparasitics. Besides these, there are multi-methods for the determination of various substances of the same group, e.g. for sulfonamides, benzimidazol anthelmintics or coccidiostats.

For drug residue analysis, most multi-methods are based on HPLC with UV detection. The advantage of HPLC in comparison with GC is reduced sample preparation: The extracted drugs can be analyzed directly without time consuming and cumbersome derivatization steps. The advantage of HPLC versus thin layer chromatography is improved separation.

Comprehensive multi-methods applying HPLC and UV-detection were developed by Petz for chloramphenicol, furazolidone and 5 sulfonamides (4); by Parks for sulfa drugs and dinitrobenzamide coccidiostats (5); by Parks for 6 nitro-containing drugs (2 nitrofuranes, sulfanitran, aklomide, zoalene and nitromide) (6); by MacIntosh for carbadox, desoxycarbadox and nitrofurazones (7); by Nose for nitrofuranes, sulfonamides and oxolinic, nalidixic and piromidic acid (8, 15); by Hori for sulfa drugs, clopidol, dinitolamide, ethopabate, nicarbazin and pyrimethamine (9) and by Malisch for numerous antimicrobials and antiparasitics with detailed demonstrations for 27 sulfonamides (10), for 8 nitrofuranes and nicarbazin (11), for chloramphenciol and meticlorpindol (12), with additional information for confirmation of results and improvement of sensitivity (using sulfanilamide, sulfaguanidine and their respective acetylmetabolites as example) (13) and with additional and detailed description of the multiwavelength-detection and spectra-confirmation for sulfapyridine, N4-acetylsulfapyridine, ethopabat, chloramphenicol, meticlorpindol, metronidazol, ipronidazol, furazolidone and nicarbazin (14).

It is by far easier to develop methods for multi-residue analysis of the same group, as the number of publications demonstrates. Then different detectors can be used, too. Examples are sulfonamides (16-31), benzimidazol anthelmintics (32-38), coccidiostats (39) and histomonostats (40). Besides these, there are numerous methods for the detection of a single compound.

1.3 CRITERIA FOR ANALYTICAL METHODS

The European Community laid down criteria for the detection of analytes in test samples for routine methods (commission's decision 87/410/EEC for substances having a hormonal or thyrostatic action. However, in the meantime they are regarded as general requirements for routine methods) and for reference methods (89/610/EEC). Similar criteria are under discussion within the Codex Committee on Residues of Veterinary Drugs in Foods. Thus, any developer of analytical methods should demonstrate the ability of the method to meet all requirements. The most important criteria of the decision 87/410/EEC are:

- accuracy: In the case of repeated analysis of the reference sample, the deviation of the mean from the true value, expressed as a percentage of the true value, shall not lie outside the following limits:

	limits
true value up to 1 μg/kg:	- 50 to + 20 %
true value over 1 and up to 10 μg/kg:	- 30 to + 10 %
true value over 10 μg/kg:	- 20 to + 10 %

- repeatability: In the case of repeated analysis of the reference sample, the coefficient of variation (C.V.) of the mean shall not exceed the following values:

	C.V.
mean up to 1 μg/kg:	0.30
mean over 1 and up to 10 μg/kg:	0.20
mean over 10 μg/kg:	0.15

- limit of detection: is equal to the mean of the measured content of representative blank samples (n>20) plus three times the standard deviation of the mean

- limit of decision: is equal to the mean of the measured content of the representative blank samples plus six times the standard deviation of the mean

Multi-methods can be used as confirmatory (mainly GC/MS), quantitative (e.g. HPLC or GC) or screening (e.g. HPLC, TLC or GC) methods. Up to now, the EEC did not lay down specific criteria for screening methods. It is difficult to develop multi-methods, which meet all the statistical criteria for quantitative methods for all compounds in exactly the same way as "single-methods": These can be optimized for each single compound. The advantage of multi-methods in the daily routine is so overwhelming that minor statistical disadvantages in comparison to "single-methods" seem to be acceptable. In any case, multi-methods can prove to meet the statistical requirements for routine methods for most substances and for screening methods for other substances.

2. MULTI-METHOD ACCORDING TO MALISCH

2.1 PROCEDURE APPLIED UNTIL 1990

Figure 1 depicts the principle of the general procedure which was applied until major changes in 1990. Residues are extracted with acetonitrile. The advantage of acetonitrile in comparison to other solvents is based on its miscibility:

acetonitrile is miscible with water, but not with solutions of salt in water. Therefore pharmacologically active substances can be extracted from water-containing food with acetonitrile in homogeneous phase. Afterwards, the co-extracted water can be separated easily by addition of sodium chloride. The organic phase is purified in miniaturized steps and concentrated to a few microliters (10 - 14).

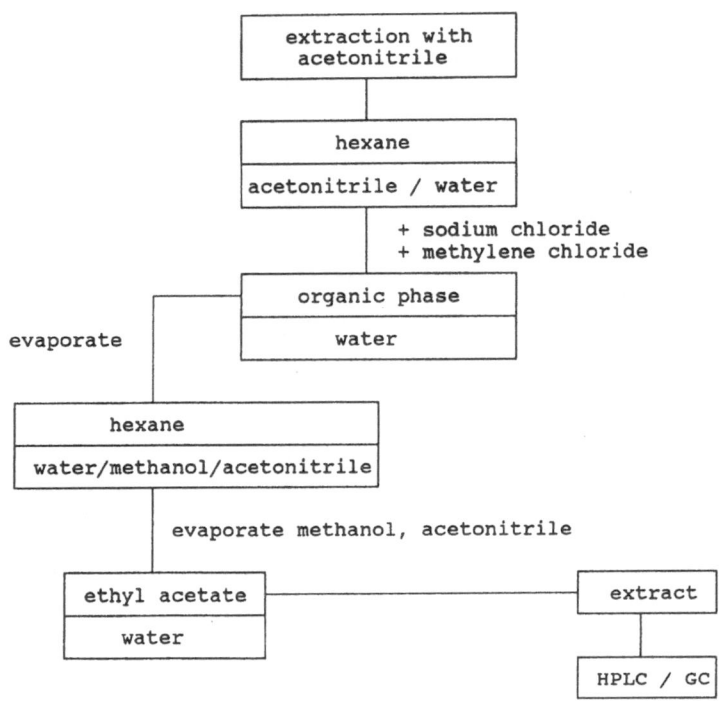

Basic detection: HPLC with UV detection
C 18-column, eluent: buffer pH 4.8 / acetonitrile
detection at 275, 315 and 360 nm

Only chloramphenicol: GC / ECD

Figure 1. General procedure of the multi-method until 1990

2.2 HPLC WITH MULTI-SIGNAL DETECTION

9 drugs were selected as representative examples for the demonstration of the multi-signal detection (see lit. 14). These compounds have different UV absorbance maxima in a range of 270 to 360 nm. Therefore, in order to analyse all the individual compounds at their highest sensitivity, three different wavelengths are selected for the chromatographic signals: 275, 315 and 360 nm (figure 2).

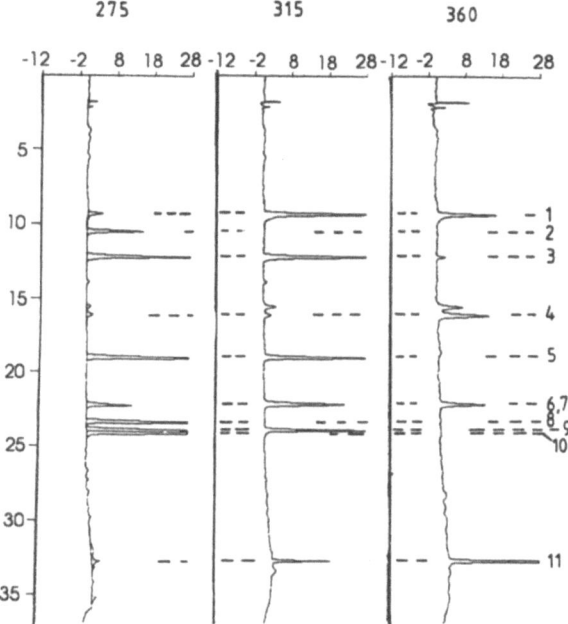

Figure 2. Chromatogram of a standard solution at different wavelengths. 1 Metronidazol (0.05 mg/kg), 2 Meticlorpindol (0.05 mg/kg), 3 Sulfapyridine (0.1 mg/kg), 4 Furazolidone (0.02 mg/kg), 5 Pyrazon (internal standard, 0.1 mg/kg), 6 Ipronidazol (0.05 mg/kg), 7 Chloramphenicol (0.05 mg/kg), 8 N4-acetylsulfapyridine (0.1 mg/kg), 9 Ethopabat (0.1 mg/kg), 10 Chloridazon (internal standard, 0.1 mg/kg), 11 Nicarbazin (0.02 mg/kg)

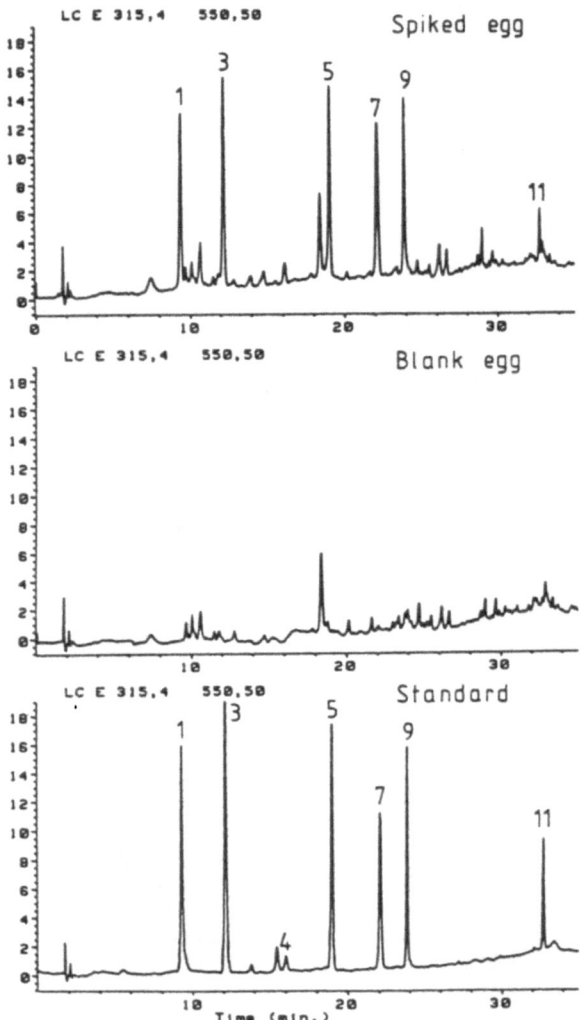

Figure 3. Chromatogram of the standard (shown in figure 2), a blank and a spiked egg sample at 315 nm.

Figure 3 to 5 show a comparison of chromatograms of spiked and blank egg extracts and of the standard mixture acquired at 275 nm (figure 4), 315 nm (figure 3) and 360 nm (figure 5). The attenuation at 315 nm was a factor of 2 and at 360 nm a factor of 7 lower than at 275 nm. These chromatograms demonstrate that the drugs of interest are found with good recoveries and detected without disturbance and with good sensitivity. The selectivity of the chosen wavelengths becomes obvious.

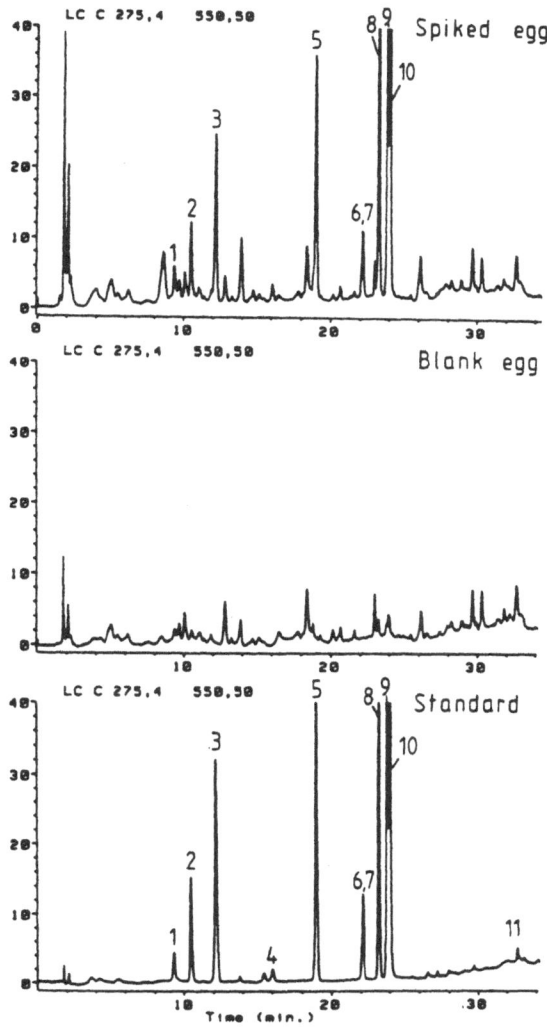

Figure 4. Chromatograms of the standard, a blank and a spiked egg sample at 275 nm.

2.3 On-line confirmation with UV-spectra

In chromatographic routine analysis, peaks are generally identified by comparing the retention times with those of a standard. Especially in trace analysis, an additional method for a peak confirmation is required. Some information on the

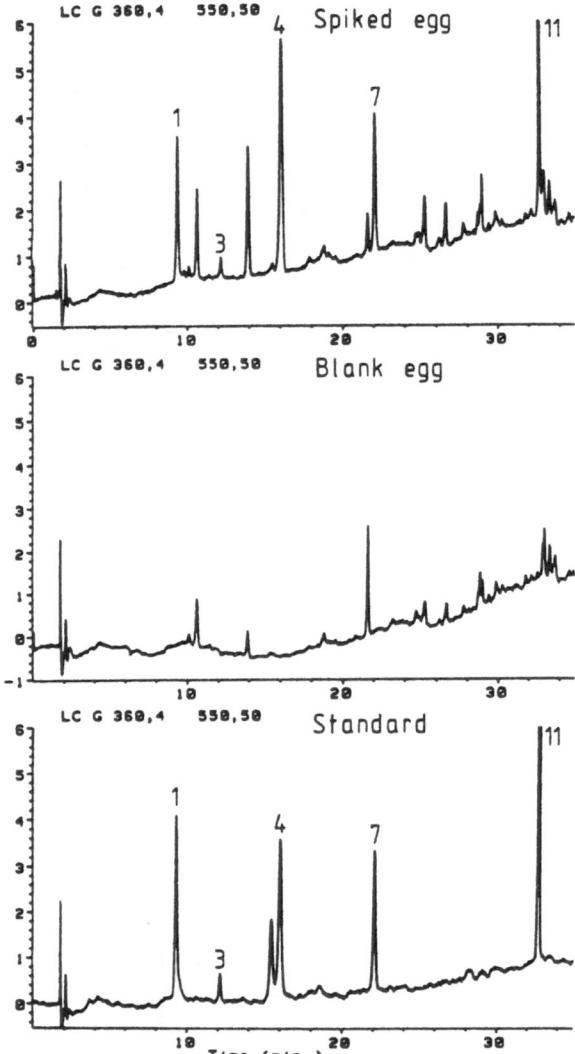

Figure 5. Chromatograms of the standard, a blank and a spiked egg sample at 360 nm.

chemical structure of a compound can be obtained from UV/Vis
spectra. With fast-scanning diode array detectors, spectra
can be acquired during a peak's elution within a few milli-
seconds. Spectra from unknown compounds can be overlayed with
those of a standard to confirm a peak's identity. Applications and limitations of this technique were demonstrated by
Malisch (10-14, with [14] as particularly detailed).

It must be mentioned that 2 different problems may arise:

- The lower the amount of a certain substance, the noisier the
  spectra. There are several possibilities to improve the
  reliability of spectra, particularly by injecting larger
  amounts of the sample or smoothing programs (see lit. 11, 13
  and 14).

- The influence of an unknown substance - for example of
  natural origin - eluting by chance at the same retention
  time as a drug must be considered. For possibilities of
  checking the spectra for co-eluting impurities by norma-
  lizing and overlaying spectra in the upslope, apex and
  downslope of a peak, see lit. 11, 13 and 14.

The EEC specified quality criteria for the detection of
residues of drugs in the commission's decision 87/410/EEC, as
mentioned in chapter 1.3. For the identification of an analyte
by HPLC with spectrometry, e.g. diode array detector, the
following criteria are demanded:

- The maximum absorption wavelength in the UV spectrum of the
  analyte should be the same as that of the standard material
  within a margin determined by the resolution of the
  detection system. For diode array detector detection, this
  is typically within +/- 2 nm.

- The spectrum of the analyte should not be visually
  different from the spectrum of the standard material for
  those parts of the two spectra with a relative absorbance
  above 10 %. This criterion is met when the same maxima are
  present and at no observed point is the difference between
  the two spectra more than 10 % of the absorbance of the
  standard material.

- For identification, co-chromatography in the HPLC step is
  mandatory. As a result, the peak presumed to be due to the
  analyte is the only one intensified.

Figure 6 shows comparisons between sample and standard
spectra of ethopabate, sulfamethazine and chloramphenicol. For
demonstration purposes, an egg sample was spiked with 0.09
mg/kg ethopabate, 0.09 mg/kg sulfamethazine and 0.05 mg/kg
chloramphenicol. Extraction and clean up followed the multi-
method. Due to the lower amounts, the spectrum of chloramphenicol becomes somewhat noisier. The spectrum of ethopabate
meets the EEC criteria perfectly, sulfamethazine approximately
(no difference in the absorption maximum and a visually different spectrum in the range 240 to 255 nm of less than 10 % of
the absorbance of the standard material) and chloramphenicol
does not meet the requirements (no difference of more than 2
nm in the absorption maximum, but a difference of more than 10
% of the absorbance between standard and sample material).

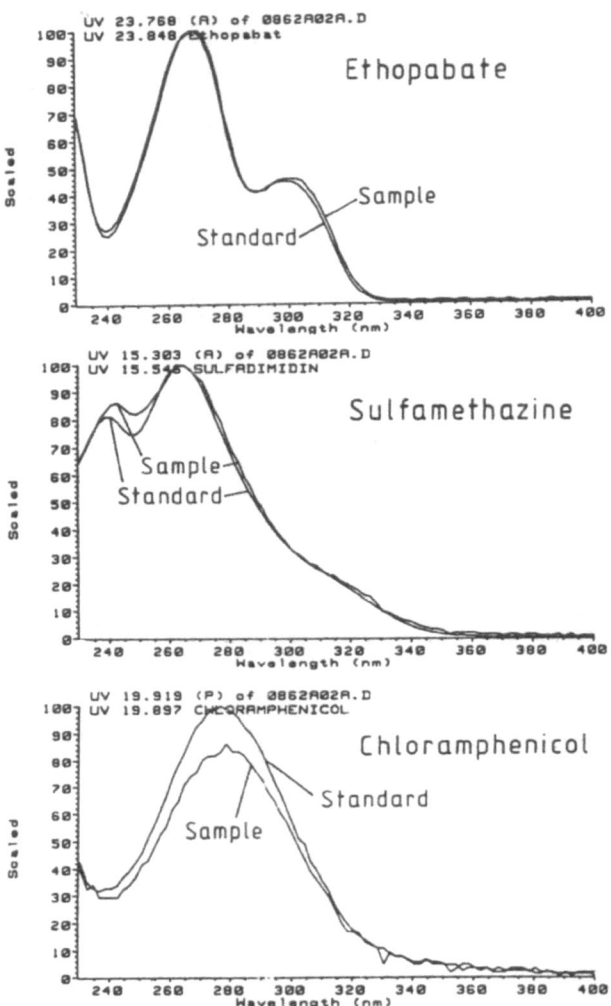

Figure 6. Overlay of spectra of spiked drugs and their respective standard spectrum (ethopabate 0.09 mg/kg, sulfamethazine 0.09 mg/kg, chloramphenicol 0.05 mg/kg)

Peaks from the sample matrix may interfere chromatographically, espacially when analyzing drugs in the range below 0.1 mg/kg. Interfering peaks do not cause false-positive but false-negative results, because the spectrum of a suspected substance in an extract becomes different from the corresponding spectrum of a drug. If the retention time and spectra are exactly the same, so far we have not made a wrong positive identification. Anyhow it is generally recommended to confirm a peak's identity with another independent method.

2.4 Collaborative tests

It is difficult to evaluate previous developed methods with statistical requirements and tolerance levels which are presented afterwards: The multi-method was developed until 1986 and published in 1986/1987. In 1987, the European Community published the statistical criteria for the first time, and final tolerances have not been fixed up to now.

2.4.1 Nitrofuranes and nicarbazin in eggs

In 1987 a collaborative test on the determination of 4 nitrofuranes and nicarbazin in eggs was conducted by the Association of German chemists. Each participant received 4 lyophilized egg samples. One blank sample and 3 samples with different concentrations were dispatched (table 1). The limit of determination should be below 5 μg/kg. Therefore, the spiked concentrations ranged from 5 μg/kg to roughly 100 μg/kg. Of course, the participants didn't know the combination of drugs in the individual samples nor the blank sample. They only knew the name of the substances to be determined.

Table 1. Spiked samples for the collaborative study "nitrofuranes and nicarbazin in eggs"

Sample	Spiked	μg/kg
A	-	-
B	Furazolidone	5
	Nitrofurazone	34
	Nicarbazin	13
C	Nitrofurantoin	9
	Nicarbazin	87
D	Furazolidone	22
	Nitrofurazone	8
	Furaltadone	28
	Nicarbazin	53

8 laboratories submitted results (table 2, see 41). All residues were detected correctly by all laboratories. There were neither false-positive nor false-negative results. The collaborative study confirmed the statistical data published for nitrofuranes and nicarbazin (11). The most important nitrofurane furazolidone and the important coccidiostat nicarbazin have recoveries of roughly 80 %. Nitrofurantoine is in the range of 75 %, nitrofurazone of 65 % and furaltadone - depending on the pH-value, which is not the best for the extraction of furaltadone - of 55 %. The commission's decision 87/410/EEC requires 70 % recovery in the range 1 to 10 μg/kg. This criterion is met by furazolidone, nicarbazin and nitrofurantoine perfectly, for nitrofurazone nearly. The method is suitable as screening procedure for furaltadone as well.

As a result, the method is able to enforce a tolerance of 0.01 mg/kg for nitrofuranes and nicarbazin in eggs, for furazolidone even of 0.005 mg/kg. (This is the tolerance which is being discussed in the European Community.) Thus, the multi-method was accepted as official method in Germany for this purpose (41).

Table 2. Results ($\mu f/kg$) of the collaborative study "nitro-furanes and nicarbazin in eggs" (8 laboratories, 8 replicates of each sample; r repeatability [intra-laboratory], R repro-duciblity [inter-laboratory])

substance	mean	r	R
Furazolidone	4.09	1.53	2.54
	16.99	4.80	11.72
Nitrofurantoine	6.89	2.01	5.48
Nitrofurazone	5.00	1.59	4.68
	22.12	5.27	16.59
Furaltadone	14.93	9.13	14.68
Nicarbazin	10.36	4.79	8.09
	42.63	26.42	44.15
	66.19	34.32	67.48

2.4.2 Chloramphenicol in milk

In 1986, the 4 state institutes (Freiburg/Offenburg, Karlsruhe, Sigmaringen and Stuttgart) and 2 community institutes (Pforzheim and Stuttgart) for chemical analysis of food in the federal state Baden-Württemberg conducted a collaborative study for the determination of chloramphenicol in milk in order to test the enforcement of the tolerance of 1 μg/kg. Table 3 presents the samples to be analysed: 4 milk samples in the range 0.5 to 7 μg/kg and 1 blank sample. The samples were dosed, that means they were obtained from animals which were treated with the analyte. The Federal Health Institute, Berlin, had provided the samples and had confirmed the "true" amount of chloramphenicol-residue by means of the RIA.

All samples were analysed correctly, that means there were neither false-positive nor false-negative results. The correspondence between the mean of results and the RIA confirmation method was very good. Each laboratory could determine residues above and below the tolerance level (1 μg/kg) reliably. Without elimination of outliers, the reproducibility in the most interesting range of about 0.6 and 1.4 ppb was below the mean of results (R = 1.29 for mean of 1.49, R = 0.34 for mean of 0.56). This data emphasizes the ability of the multi-method to enforce a tolerance of 1 μg/kg chloramphenicol in milk.

Table 3. Collaborative study "chloramphenicol in milk"
(6 laboratories, 8 replicates per sample, results in μg/kg)

sample no.	confirmed amount (RIA)	mean of results	Detailed results
1	5.88	6.82	3.93, 5.39, 5.76, 7.52, 8.06, 10.24
2	3.06	2.95	1.58, 2.56, 2.75, 3.21, 3.54, 4.14
3	1.28	1.49	1.18, 1.25, 1.29, 1.44, 1.67, 2.20
4	0.57	0.50	0.38, 0.49, 0.51, 0.53, 0.63, -
5	<0.1	<0.1	0.00, 0.00, 0.00, 0.00, 0.10, 0.10

5 laboratories submitted results. Table 5 lists the results. In the left column, the results from our laboratory are summarized, in the right column the results from other collaborators. On the whole, the results were more or less similar. Most laboratories determined in sample 1 sulfamethazine as the most important residue in the range 2 to 3 mg/kg, then sulfamerazine with about 300 μg/kg and last sulfadiazine at about 100 μg/kg. Sulfathiazole was not detected by any laboratory. This could be due to a quicker metabolism of this substance.

In the second sample, sulfapyridine and sulfanilamide were determined in roughly the same concentration of 200 to 300 μg/kg. Sulfaguanidine was not detected, besides by a single laboratory which had applied a specific method for the sensitive detection of sulfanilamide and sulfaguanidine.

2.4.3 Sulfonamides in meat

In 1986, the German Federal Health Institute conducted a preliminary study for the determination of sulfonamides in dosed meat samples. The objective was the comparison of different methods which were established in different laboratories. For this, two swine were treated with 2 different drugs containing different sulfonamides (table 4). The collaborators knew only that the samples were dosed with sulfonamides, but not with which type of the nearly 30 different sulfonamides.

Table 4. Treatment of swine for the preliminary study "sulfonamides in meat"

1. With "Poly-Sulfa-Komplex":
   sulfamerazine (108 mg/ml), sulfadiazine (87 mg/ml),
   sulfamethazine (86 mg/ml), sulfathiazole (48 mg/ml)

2. With "Terpoleucin":
   sulfanilamide (105 mg/ml), sulfaguanidine (30 mg/ml),
   sulfapyridine (15 mg/ml)

Table 5. Mean of the results (μg/kg) of the determination of sulfonamides in dosed meat samples. 8 Replicates for each sample (in brackets: coefficient of variation in %)

Sample 1	multi-method	(CV)	other collaborators			
sulfamethazine	3106	(14)	1500,	2750,	2970,	nd
sulfamerazine	370	(11)	340,	215,	340,	370
sulfadiazine	95	(15)	83,	190,	64,	nd
sulfathiazole	nd		nd,	nd,	nd,	nd
Sample 2						
sulfanilamide	206	(11)	270,	370,	58,	nd
sulfaguanidine	nd		50,	nd,	nd,	nd
sulfapyridine	297	(9)	258,	116,	211,	nd

Figure 7 depicts the chromatogram of the extract of sample no. 1. Dominant peaks are sulfadiazine, sulfamerazine and sulfamethazine. With regard to these substances, only a negligible amount of sulfathiazole was detectable. Even in traces, sulfanilamide (as metabolite or impurity) was not detectable. Pyrazon (= chloridazon) and benzthiazuron were used as internal standards.

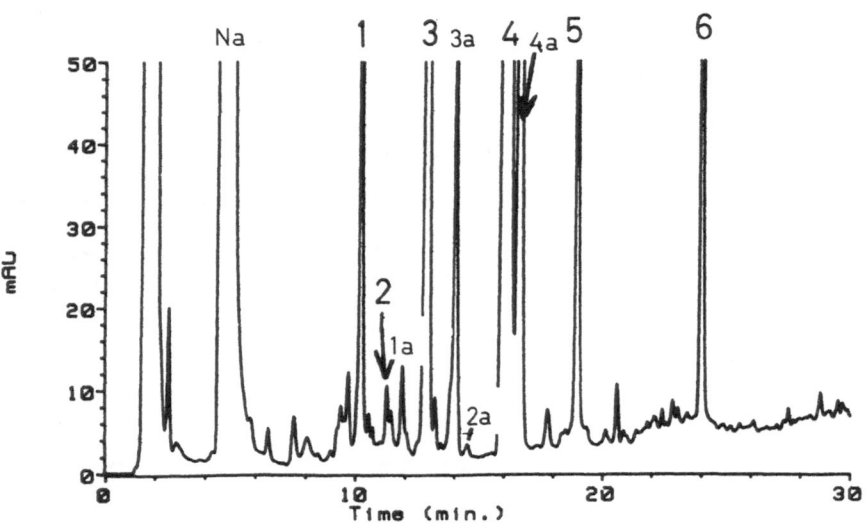

Figure 7. Chromatogram of a dosed meat sample (Na nicotinamide, 1 sulfadiazine, 1 a N4-acetalsulfadiazine, 2 sulfathiazole, 2a N4-acetylsulfathiazole, 3 sulfamerazine, 3 a N4-acetylsulfamerazine, 4 sulfamethazine, 4 a N4-acetylsulfamethazine, 5 pyrazon, 6 benzthiazuron)

As mentioned, the samples were obtained from swine which were treated with sulfonamides. Therefore we tried to identify metabolites in the extract. In pigs the acetylation pathway is known to be predominant. Although N4-acetylmetabolites were not available in 1986, we looked for these substances in our chromatograms. With respect to retention time and UV spectra, we assumed that 2 major peaks presented N4-acetylsulfamerazine and N4-acetysulfamethazine. In the meantime, we got certain N4-acetylderivates and could confirm our results. Further, we could identify smaller peaks as N4-acetylsulfadiazine and N4-acetylsulfathiazole.

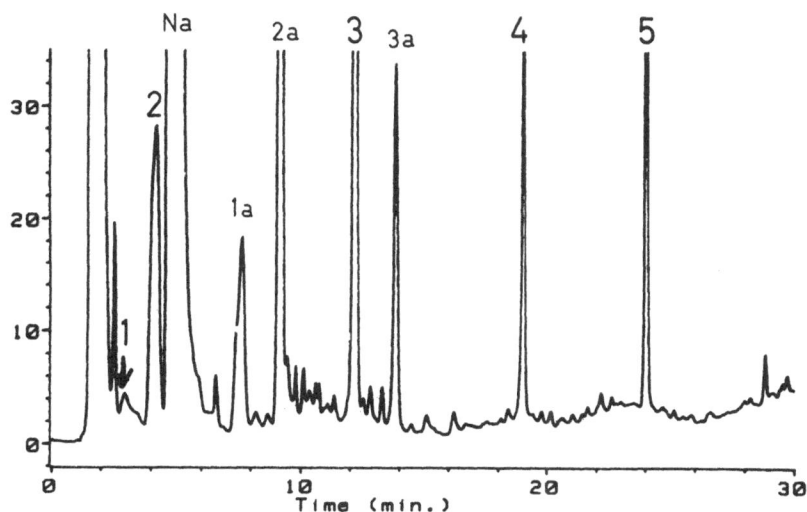

Figure 8. Chromatogram of a dosed meat sample (1 sulfa-guanidine, 1a N4-acetylsulfaguanidine, 2 sulfanilamide, 2a N4-acetylsulfanilamide, 3 sulfapyridine, 3a N4-acetyl-sulfapyridine, 4 pyrazon, 5 benzthiazuron)

Figure 8 depicts the chromatogram of the extract of sample no. 2. Sulfanilamide and sulfapyridine were detected, as well as their acetylmetabolites. Sulfaguanidine was not detectable, but its acetylmetabolite.

As a result, less polar sulfonamides - like sulfametha-zine or sulfadiazine - can be detected in meat samples reli-ably. The polar sulfanilamide and especially sulfaguanidine cause more problems. Their detection depends on the amount of their respective residue (limit of detection 0.1 mg/kg for sulfanilamide, 0.5 mg/kg for sulfaguanidine). With the multi-method also N4-acetylmetabolites are detectable. The recov-eries are roughly the same or even better than the recoveries of the respective sulfonamides. The more important sulfon-amides as sulfamethazine or sulfadiazine can be determined in the range from 30 µg/kg to 3000 µg/kg.

2.4.4 Chloramphenicol in meat

In 1989, the German Federal Health Institute conducted a collaborative study for the determination of chloramphenicol in dosed meat samples. The procedure was described in detail (12) with the exception of an important change for an improved stability of the silylated extract: When the extract is blown down after silylation, the residue is re-solved in hexane with 5 % silylation agent (pyridin/TMCS/HMDS) instead of solving in pure hexane, only. If extracts are to be diluted, this has to be done with the same mixture (hexane with 5 % silylation agent).

15 laboratories submitted results. From these, 3 laboratories changed the derivatization procedure considerably (use of a different silylation agent, wrong dilution of extracts or blowing down of the silylation agent in a water bath). 2 laboratories were eliminated as outliers with respect to blank samples (almost the same amount in blank sample as in dosed meat). The remaining 10 laboratories had good results: "True" value in the dosed meat sample was 11 ppb chloramphenicol, determined by the Federal Health Institut applying RIA. The mean of the 10 laboratories was 9.48 ppb (=86.2 % recovery) without correction by means of internal standards. The reproducibility R was 6.55 ppb. The blanks were almost generally below 1 ppb, thus less than 10 % of the determined residue.

Table 6. Confirmation of chloramphenicol in dosed meat samples on 2 columns (SE-30 resp. DB-17), with 3 internal standards. (Chemische Landesuntersuchungsanstalt Freiburg, 1989, as part of a collaborative study with dosed meat [11 ppb as "true" value], mean of 8 replicates).

| | SE-30 | | | DB-17 | | |
	PCB-153	metaCA	MCCA	PCB-153	metaCA	MCCA
mean	11.6	9.34	7.37	9.24	11.9	11.2
CV	11.9	8.4	14.8	10.7	13.0	23.7

	all determinations (2 columns, 6 int.std)	mean of 2 columns for: PCB-153	metaCA	MCCA
mean	10.1	10.4	10.9	9.3
CV	8.7	9.7	7.7	19.6

Our laboratory had confirmed the results on different GC columns and with respect to different internal standards. PCB 153 is an internal standard which is added after the clean up-procedure before the injection and corrects mainly the unreproducible manual injection (or controls the succesful injection with an autosampler). Meta-Chloramphenicol (meta-CA) and Monochlor-Chloramphenicol (MCCA) would be "true" internal standards, because these are added before the extraction and correct any losses in the whole procedure. Anyhow, meta-CA gives better results than MCCA. These comprehensive tests confirm the suitability of the internal standards as well as the agreement of the results on different columns (table 6).

2.4.5 Sulfonamid standard solutions

As preparation for another collaborative study for the determination of sulfonamides in meat, in 1991 we organized an international ringtest to evaluate the comparability of standard solutions. 2 standard solutions were to be analyzed, with 6 sulfonamides and 4 N4-acetylmetabolites in solution 1 and 8 sulfonamides in solution 2. Each participant knew the names of 12 sulfonamides which could have been added, plus the major N4-acetylmetabolites. The concentration range was 1 to 100 ng/μl. Thus, it was a qualitative and quantitative test.

26 laboratories submitted results (table 7) (44). The very most results were neither false-positive nor false-negative. The mean of all laboratories is very close to the "true" value which was prepared in our laboratory. Anyhow, the most important result is the relatively bad precision: Without elimination of laboratories, the reproducibilty R is in the range 50 to 90 % for sulfonamides and 70 to 140 % for acetylmetabolites, and this without any "disturbing" influence of a clean up procedure. Thus, if a method is to be tested in a collaborative study, a calibration solution with the substances of interest must be supplied to avoid a false interpretation of the tested method (unsufficient precision is caused not only by the method, but by the calibration solutions of the participants, as well).

Table 7. Results of a collaborative study with standard solutions

	soln. no.	"true" value	mean of labs	R (%)	number of laborat.
S.anilamide	1	10.1	10.5	82.7	24
S.anilamide	2	5.0	6.1	111.8	21
Acet.s.anilamide	1	10.1	10.7	67.9	12
S.guanidine	2	16.0	16.6	81.4	21
S.diazine	1	103.0	92.5	60.6	25
S.diazine	2	2.1	2.2	90.8	22
Acet.s.diazine	1	11.0	12.9	92.5	8
S.thiazole	1	5.0	5.4	53.8	17
S.thiazole	2	40.0	44.2	66.6	23
S.pyridine	1	60.0	59.4	58.5	21
S.merazine	2	20.0	20.1	41.6	26
Acet.s.merazine	1	10.5	9.9	90.0	11
S.methazine	1	5.0	5.9	116.3	24
S.methazine	2	10.0	9.9	37.4	24
Acet.s.methazine	1	10.4	16.9	144.3	11
S.quinoxaline	1	5.0	4.9	52.8	19
S.methoxazole	2	100.0	100.2	26.4	22
S.dimethoxine	2	3.2	3.4	47.4	18

Table 8. Results from state institutes for chemical analysis of food, Baden-Württemberg, 1984-1990, with application of the multi-method

Eggs and egg products

Number of analyzed samples: 4,380

Number of samples with residues:
143 Chloramphenicol	(	0.1 -	37	μg/kg)
135 Nicarbazin	(	2.0 -	5200	μg/kg)
231 Clopidol	(	1.3 -	730	μg/kg)
3 Furazolidone	(	29 -	49	μg/kg)
1 Sulfadiazine	(	20		μg/kg)
1 Sulfachlorpyrazine	(	580		μg/kg)
2 Dimetridazole	(	7 -	26	μg/kg)
1 Ronidazole	(	38		μg/kg)

Poultry

Number of analyzed samples: 1258

Number of samples with residues:
188 Clopidol	(	1.4 -	2110	μg/kg)
109 Nicarbazin	(	3.0 -	2700	μg/kg)
7 Chloramphenicol	(	0.5 -	35	μg/kg)
2 Sulfaquinoxaline	(	34 -	110	μg/kg)
1 Ronidazole	(	31		μg/kg)

Meat and meat products

Number of analyzed samples: 2814

Number of samples with residues:
56 Chloramphenicol	(	0.5 -	13000	μg/kg)
21 Sulfamethazine	(	9.0 -	520	μg/kg)
3 Sulfadiazine	(	60 -	190	μg/kg)
2 Sulfamerazine	(150	-	780	μg/kg)
2 Sulfamethoxypyridazine	(160	-	22000	μg/kg)
3 N4-acetylsulfonamides	(	2.5 -	970	μg/kg)
3 Clopidol	(	23.3 -	100	μg/kg)

Fish and fish products

Number of analyzed samples: 1732

Number of samples with residues:
70 Chloramphenicol	(	0.1 -	86	μg/kg)
1 Sulfamerazine	(	1400		μg/kg)
1 Sulfamethazine	(	120		μg/kg)
1 N4-acetylsulfamethazine	(	80		μg/kg)
(251 Malachite green)	(	0.6 -	2400	μg/kg)

Milk and milk products

Number of analyzed samples: 2710

Number of samples with residues:
66 Chloramphenicol	(	0.2 -	7	μg/kg)

2.5 Results from the official foodstuff supervision

A few results from the state institutes for chemical analysis of food in the federal state Baden-Württemberg (42) demonstrate the efficiency and practicability of the multi-method (table 8). From 1984 to 1990, these institutes analyzed about 13,000 samples by means of the multi-method. It should be mentioned that some substances - as nicarbazin and clopidol - were not considered in the beginning of the development of the multimethod. Applying the method to food, we sometimes observed unidentified peaks. We could identify clopidol and nicarbazin as unknown peaks. This was the first detection of these coccidiostats in the official foodstuff surveillance in Germany. In the same way, malachite green was found out as residue in fish samples: The extracts of the samples, which were analyzed according to the multi-method, were blue-green. The state institute in Sigmaringen could identify malachite green causing this colour. Later, a specific method was developed for the determination of malachite green-residues (45 - 47).

2.6 Procedure developed in 1990

We tried numerous attempts to simplify the procedure and at the same time to meet the strict statistical requirements of 1987 laid down in the decision number 410 of the commission of the European Community. The analytical details will be published soon (48), thus we will give a short overview of the most important changes:

- optimization of the extraction step: With meat and eggs as sample material, we tested different techniques (ULTRA-TURRAX high-efficiency dispersing, WARING BLENDOR extractor, shaking with a horizontal shaker for the treatment of 4 to 8 samples at the same time, ultrasonic treatment with a ultrasonic bath and with a LABSONIC homogenizer [with a power output in excess of 340 Watts]).

As a result, shaking of homogenized samples proves to be an efficient way of extraction, above all for a series of 8 samples at the same time, for instance. These samples can be extracted in screw-cap tubes fixed in circular or horizontol shakers without any danger of cross-contamination by shafts, stirring rods, probes or other devices. As the data presented later in this presentation will document, it is possible to extract about 90 - 100 % of sulfonamides, chloramphenicol and some pretested antiparasitics with a single extraction step out of meat samples.

- Composition of the extraction solvent: In former times, meat, milk and eggs were extracted directly with acetonitrile. We found that recoveries especially for sulfonamides in meat could be improved considerably, if the meat sample was extracted first with double the amount of water and if acetonitrile was added to the meat-water homogenate.

If meat is extracted directly with acetonitrile, the meat particles form a coagulate, and it is difficult to extract substances out of the inner flocculate with its relatively firm shape. If meat is extracted first with buffer by shaking and afterwards acetonitrile is added, the fine meat matrix particles sink to the bottom of the tube, somehow homogenous in a water-acetonitrile-extract.

- Renunciation of any halogenated solvents: As in the previous form, the coextracted water must be removed. Usually this is done by the addition of salt and methylene chloride. With respect to the toxicity of halogenated solvents and to the environmental effects, the analyst should avoid the use of any halogenated solvents. Further, German laboratories have severe problems discharging the waste of halogenated solvents. For example, the state institute for chemical analysis of food in Stuttgart was about to stop its analysis in general with respect to full halogenated waste tanks in January 1990.

We could replace methylene chloride succesfully by butyl-methylether/hexane.

- Addition of a keeper for the evaporation process: We found out that co-extractives especially from meat samples may absorb a small amount of sulfonamides if the extracts are evaporated to complete dryness. Although sulfonamides can be concentrated as pure solutions in acetonitrile or methanol without any losses, sulfonamides can be partly bound to coextractives. In order to avoid total evaporation, a solvent with a significant higher boiling point can be added as "keeper", that means it keeps the substances of interest in solution. For our purposes, ethylene glycol proved to be a good keeper. For the concentration of the crude extract, we add the ethylene glycol to the organic phase and evaporate down to "ethylene glycol"-dryness. This reduces the danger of binding to coextractives.

- Automatization of the evaporation process: We could help to improve the technique of centrifugal vacuum concentration considerably (43).

Centrifugal vacuum concentrators can be applied in residue analysis succesfully: They prevent the samples from bumping and foaming. Further, they can automate the evaporation process. In comparison to rotary evaporators, they save time for the laboratory staff when running series of 4 or more samples.

Two important factors influence the speed of any evaporation process: vapor pressure and supply of energy for continuous evaporation. For a rotary evaporator, the vapour pressure can be controlled by a vacuum controller and the energy by the temperature of the waterbath.

A centrifugal vacuum concentrator obeys the same principles. However, whereas the control of the vapour pressure is no problem at all, the supply and control of the energy of evaporation was a severe problem. Up to now, commercially available vacuum concentrators as the SPEED VAC CONCENTRATOR use a chamber heater (controlled at 45 °C, for instance) located in the inner centrifuge wall. In order to evaporate solvents, a very low vacuum is applied (0.01 mbar, for example), and the evaporating solvents are trapped in a refrigerated condensation trap.

The main disadvantage of such a system is the lack of energy supply for a continuous evaporation process: After some minutes, the liquids are cooled down to low temperatures (for acetonitrile to about - 10 °C). Therefore, the evaporation process slows down considerably. Because of the high vacuum, there are no gas particles left for an energy transport from the heated inner side of the centrifuge tubes, and the radiation of heat from a surface of 45 °C is minimal. Thus, it is unacceptably time consuming to evaporate 8 samples each of 100 ml extract.

An approach to provide more energy was the introduction of a heater built in the side wall of the chamber to heat the chamber up to 80 °C. Unfortunately, this didn't solve the principal problem: If organic extracts are preheated to higher temperatures, for example 60 °C, they tend to bump a few seconds after the vacuum is turned on, and a few minutes later the solvents are cooled down again.

A more sophisticated approach to provide sufficient energy was the introduction of IR radiation. Two IR lamps (controlled by a thyristor) were located above the glass safety cover. After a couple of days, the user gets some experience in the optimal operation range of the thyrister control.

The advantage is by far improved continuous energy transfer and thus a considerable acceleration of the evaporation process. However, two important disadvantages remain: First, it is very difficult to estimate the temperature on the centrifuge tubes (as a result, overheating of the samples is possible, even bumping of the extracts), and second, the rotor beakers must be adjusted to the glass tubes by means of filling with chips of aluminium foil. Otherwise the gap between the centrifuge tubes and the rotor is too big and diminishes the energy transfer considerably, especially to the glass tube bottom.

The newly developed CHRIST centrifugal vacuum concentrator has 4 IR lamps integrated in the bottom of the centrifuge sides. Brown-glass tubes absorb the IR radiation directly: A special rotor takes these tubes in a way that they are radiated directly. Thus, any contact problems between the steel or aluminium beakers of the rotor and the centrifuge tubes are avoided. Additionally, a new vertical rotor with special flasks allows the drying of higher volumes on the smallest radius with a constant and maximal size of the evaporating surface during the whole evaporation time.

Further - and even more important - the temperature on the surface of a selected centrifuge tube can be checked electronically and is representative for all other tubes. This controls the IR heating system. Thus, the user can set a temperature of 40 °C, for instance, and the control panel will activate the IR lamps whenever necessary, that is in the heating-up time as well as in the evaporation time. Now, in combination with the vacuum controller, the user has a very easy possibility to control the two important parameters of the evaporation process.

Another advantage of this system is the possiblity to control the evaporation process visually: The evaporated solvents are condensed in the outlet of a special so-called "Hybrid-pump" by a high efficiency condensor. In a graduated glass bottle, the progress can be watched very simply by determination of the volume of the condensated solvent.

- Simplification of the former liquid/liquid purification steps: After evaporation of the organic solvent, the crude extract is dissolved in ethylene glycol. This raw extract is purged with hexane, acetonitrile and buffer into a smaller screw-cap tube and defatted by suction of the upper hexane layer.

  According to the previous procedure, the lower phase had to be evaporated in order to get rid of the organic solvents. Now, instead of this evaporation step, the lower ethylen glycol/water/acetonitrile-phase is diluted with another small portion of water and then extracted with ethyl acetate twice.

- Final volume: The combined ethyl acetate phases are added by a small amount of ethylene glycol and evaporated to "dryness". These extracts are filled up to 1 ml with water and ready for the HPLC analysis. In this final solution, we have a mixture of ethylene glycol and water with about 50 % of each solvent.

- Detection: Basically, the substances are detected by HPLC with diode array detection, as described in detail (14). In contrast to the former chromatography, the new ethylene glycol/water-mixture renders it possible to inject 5 times more extract onto the HPLC column than the previous used acetonitrile/methanol/water-mixture.

  So, despite the reduced sample amounts and the increased final volume, we could achieve the detection limits as published earlier.

- Chloramphenicol-Screening with immunoassay: Chloramphenicol is detected by an immunoassay as a first screening. The extracts can be diluted with water and then directly used for the Quik-Card test (49 - 50) or any other ELISA or RIA. This is by far easier and less time consuming than an additional silicagel adsorption chromatography with derivatization and GC/ECD detection. Nevertheless, ethylene glycol would disturb the silylation process and must be removed, e.g. on a C18-cartridge.

One of the main analytical performance characteristics is the recovery. Table 9 shows the recoveries for 8 sulfonamides, acetylsulfamethazine, internal standards, nicarbazin, chloramphenicol, and pretested nitrofuranes and flubendazole. It must be pointed out the fact that this is a "9-month / 8 people"-statistics, that means these are not data from an optimal sequence of 8 replicates in a single series analyzed by one person, but from 8 different people who tested the method during a 9-month period. The most data originate from a single spiked routine sample: Generally, we run a series of 8 samples. 7 samples are regular blank samples, and one of these 7 is spiked. These spiked routine samples contribute to this 9-month statistic. From these 8 people, only one had experience in this method. The other 7 people tried the modified procedure for the first time.

The recoveries are in the range 80 - 110 % for all sulfonamides, the internal standards, nicarbazin, furazolidone and flubendazol. Nitrofuranes and anthelmintics were pretested in an early stage of the method validation, thus not optimized, although with about 98 % for flubendazol there is not much space for further optimization. Up to 50 replicates were run for some substances.

These data meet all requirements of the European Community and the discussed requirements of the Codex Alimentarius Commission.

Sulfanilamide was evaluated at different wavelengths in the UV range and with fluorescence detection. Even the polar sulfanilamide and the unpolar nicarbazin give recoveries in the range 80 - 100 %.

The coefficient of variation should be less than 15 %, according to the requirements of the European Community and the discussed requirements of the Codex Alimentarius Commission. These requirements are met by all substances besides sulfanilamide, which is in the range 15 and 20 % for the evalutation with fluorescence detector or UV at 260 nm, further for all internal standards and nicarbazin, if nicarbazin is corrected by means of an internal standard.

It is not recommendable to present a method until confirmation through other laboratories. Up to now, we received data from 7 other laboratories which applied the method. Some of these data originate from an orientating ring test with spike solutions, others from internal recovery studies. These data confirm our recoveries.

For comparison, we added results from our laboratory to an 8-laboratory pretest (table 10). For our laboratory, we chose data from the 8 coworkers mentioned earlier, and these from the first application of the method. That means, that our institute contributes altogether 8 replicates, one of each of the 8 coworkers. These are "worst case"-conditions: In most cases, the developer of a method presents statistical data from one skilled person and from a single series, thus under optimal conditions.

With these data from other laboratories, we calculated the reproducibility. Generally, the reproducibility is far below the mean of the recovery. This meets the basic requirement: R < mean. Only for acetylsulfamethazine the reproducibility is in the range of about 80 % of the mean, for all other substances is the reproducibility about one third to half of the spiked amount.

AS A RESULT: The multi-method was simplified successfully. For sulfonamides and nicarbazin (marking the polar and unpolar end of the substances of interest), all strict statistical requirements of the European Community are met. Up to now, the statistical data has been confirmed by 7 other laboratories.

Table 9. Recoveries (%) of substances spiked to meat samples: sulfonamides (100 ppb), acetylsulfamethazine (50 ppb), nicarbazin (50 ppb) and chloramphenicol (100 ppb). Statistic data as mean from 8 different people and analysis during a 9-month-period in the CLUA Freiburg

	S.-diazine	S.-pyrid.	S.-thiaz.	S.-methaz. UV/275	S.-methaz. UV/260
mean	99.1	96.7	78.9	107.1	101.5
replicates	50	12	40	46	13
C_v	14.3	13.7	12.0	12.1	7.7

	S.meth-oxazole	S.di-methox.	S.-meraz.	Acetyls.-methaz.
mean	100.7	98.1	96.8	103.8
replicates	9	47	41	41
C_v	12.2	10.7	8.8	13.1

	S.anil. Fluor.	S.anil. UV/290	S.anil. UV/275	S.anil. UV/260	Nicarb.	Nicarb. [Oryz.-korr.]
mean	99.6	94.2	82.2	76.5	91.8	101.9
replicates	16	18	9	11	20	20
C_v	18.2	24.6	46.8	17.9	18.1	14.9

	CAP	Ethidim.	Benth.	Oryz.
mean	117.3	106.0	102.8	91.4
replicates	11	45	50	43
C_v	13.4	6.8	7.0	13.0

	Fura-zolid.	N.fura-zon	Nifur-sol	Fura-droxyl	Flubend.
mean	92.7	69.7	65.8	84.9	98.1
replicates	11	4	4	4	3
C_v	23.8	14.1	40.9	5.9	3.8

Table 10. 8-laboratory test: Recoveries (%) for sulfonamides (100 ppb), acetylsulfamethazine (50 - 100 ppb) and nicarbazin (50 - 100 ppb), spiked to meat samples

	S.meth-azine	S.di-methox.	S.-meraz.	S.methoxy pyrid.
mean	95.5	97.5	91.7	82.8
replicates	41	33	41	5
reproducibility	43.8	44.5	36.9	24.5

	S.anil-amide	S.-diazine	S.-pyrid.	S.-thiaz.
mean	77.6	89.4	92.4	77.7
replicates	28	42	16	41
reproducibility	31.7	32.4	42.1	45.1

	Acetyls.-meth.	Nicarb.	Nicarb. (Oryz.-korr)
mean	103.9	88.3	96.7
replicates	23	37	26
reproducibility	85.5	42.5	75.3

	Ethidim.	Benz-thiaz.	Oryzalin
mean	104.5	108.0	91.6
replicates	28	36	29
reproducibility	49.4	43.7	47.9

References

1. Somogy, A. (1982) Bundesgesundheitsbl. 25:265-370
2. Deutsche Forschungsgemeinschaft (1983) "Rückstände in Lebensmitteln tierischer Herkunft" ('Residues in Food of Animal Origin'), Mitteilung X der Kommission zur Prüfung von Rückständen in Lebensmitteln, p. 14-15, Verlag Chemie, Weinheim, Germany
3. Petrausch, R. (1983) "Delta-Index", Delta medizinische Verlagsgesellschaft, D-1000 Berlin 51 (West), Germany
4. Petz, M. (1983) Z. Lebensm. Unters. Forsch. 176:289-293
5. Parks, O.W. (1985) J. Assoc. Off. Anal. Chem. 68:20-23
6. Parks, O.W. (1989) J. Assoc. Off. Anal. Chem. 72:567-569
7. MacIntosh, A.E. and G.A. Neville (1984) J. Assoc. Off. Anal. Chem. 67:958-962
8. Nose, N. et al. (1987) J. Assoc. Off. Anal. Chem 70:714-717
9. Hori, Y. (1983) J. Food Hygienic Society of Japan 24:447-453
10. Malisch, R. (1986) Z. Lebensm. Unters. Forsch. 182:385-399
11. Malisch, R. (1986) Z. Lebensm. Unters. Forsch. 183:253-266
12. Malisch, R. (1987) Z. Lebensm. Unters. Forsch. 184:467-477
13. Malisch, R. (1987) Archiv für Lebensmittelhygiene 38:41-47
14. Malisch, R. and L. Huber (1988) J. Liquid Chromatogr. 11:2801-2827
15. Nose, N. et al. (1986) J. Assoc. Off. Anal. Chem. 70:714-717
16. Alawi, M.A. and H.A. Rüssel (1981) Fresenius Z. Anal. Chem. 307:382-384
17. Simpson, R.M. et al. (1985) J. Assoc. Off. Anal. Chem. 68:23-27
18. Thomas, M.H. et al. (1983) J. Assoc. Off. Anal. Chem. 66:881-883
19. Thomas, M.H. et al. (1983) J. Assoc. Off. Anal. Chem. 66:884-892
20. Holtmannspötter, H. and H.P. Thier (1982) Dtsch. Lebensm. Rdsch. 78:347-350
21. Haagsma, N. (1985) Z. Lebensm. Unters. Forsch. 181:45-46
22. Schlatterer, B. (1983) Z. Lebensm. Unters. Forsch. 176:20-26
23. Wyhowski de Bukanski, B. et al. (1988) Z. Lebensm. Unters. Forsch. 187:242-245
24. Aerts, M.M.L. et al. (1988) J. Chromatogr. 435:97-112
25. Rychener, M. et al. (1990) Mitt. Gebiete Lebensm. Hyg. 81:522-543
26. Matusik, J.E. et al. (1990) J. Assoc. Off. Anal. Chem. 73:529-533
27. Agarwal, V.P. (1991) "HPLC determination of 8 sulfonamides in milk", The Fourth Chemical Congress of North America, August 25-30, 1991, New York, publication in preparation
28. Long, A.R. et al. (1990) J. Agric. Food Chem. 38:423
29. Smedley, M.D. and J.D. Weber (1990) J. Assoc. Off. Anal. Chem. 73:875-879
30. Takatsuki, K. and T. Kikuchi (1990) J. Assoc. Off. Anal. Chem. 73:886-892
31. Pacciarelli, B. et al. (1991) Mitt. Gebiete Lebensm. Hyg. 82:44-55
32. Barker, S.A. et al. (1990) J. Assoc. Off. Anal. Chem. 73:22-25
33. Wilson, R.T. et al. (1991) J. Assoc. Off. Anal. Chem. 74:56-67

34. LeVan, L.W. et al. (1991) J. Assoc. Off. Anal. Chem.
    74:487-493
35. Long, A.R. et al. (1989) J. Assoc. Off. Anal. Chem.
    72:739-741
36. Long, A.R. et al. (1990) J. Assoc. Off. Anal. Chem.
    73:860-863
37. Tai, S.S.C. et al. (1990) J. Assoc. Off. Anal. Chem.
    73:368-373
38. Marti, A.M et al. (1990) J. Chromatogr. 498:145-157
39. Gallicano, K.D. et al. (1988) J. Assoc. Off. Anal. Chem.
    71:48-50
40. Aerts, R.M.L. et al. (1991) J. Assoc. Off. Anal. Chem.
    74:46-55
41. Amtliche Sammlung von Untersuchungsverfahren nach
    § 35 LMBG, Methode L 05.01/02 von Dezember 1988
42. "Aus der Arbeit der Chemischen Landesuntersuchungs-
    anstalten Baden-Württembergs", Gemeinsamer Jahresbericht,
    1984 - 1990
43. Malisch, R. et al. (1991) Fresenius J. Anal. Chem. 341:
    449-456
44. Lippold, R., B. Bourgeois and R. Malisch (1992) Lebens-
    mittelchem. Gerichtl. Chem. (publication in preparation)
45. Edelhäuser, M. and E. Klein (1986) Dtsch. Lebensm. Rdsch.
    82:386-389
46. Klein, E. and M. Edelhäuser (1988) Dtsch. Lebensm. Rdsch.
    84:77-79
47. Lippold, R., M. Edelhäuser and E. Klein (1992) Dtsch.
    Lebensm. Rdsch., in press
48. Malisch, R. B. Bourgeois and R. Lippold, publication in
    preparation
49. Nouws, J.F.M. et al. (1987) Archiv Lebensmittelhyg.
    38:7-9
50. Nouws, J.F.M. et al. (1987) Archiv Lebensmittelhyg.
    38:9-11

CONTRIBUTORS

Dr. V. K. Agarwal

The Connecticut Agricultural
Experiment Station.
Box 1106
123 Huntington Street
New Haven, CT 06504
U. S. A.

Prof. S. A. Barker

Department of Veterinary Physiology,
Pharmcology, and Toxicology
School of Veterinary Medicine
Lousiana State University
Baton Rouge,
Lousiana 70803
U. S. A.

Dr. C. J. Barnes

Office of New Animal Drug
Evaluation
Food and Drug Administration
Beltsville
MD 20705
U. S. A.

Dr. M. S. Brady

Department of Biochemistry and
Microbiology
Cook College/New Jersey
Agricultural Experiment Station
Rutger-the State University of
New Jersey
New Brunswick, N.J. 08903-0231
U. S. A.

Dr. M. C. Carson

Division of Veterinary Medical
Research
Center for Veterinary Medicine
Food and Drug Administration
Barc East, Bldg. 328-A
Beltsville, MD 20705
U. S. A.

Dr. S. E. Charm

President
Penicillin Assays Inc.
36 Franklin Street
Malden, MA 02148
U. S. A.

Dr. E. Daeseleire

Laboratory of Food Analysis
University of Ghent
Harelbekestrat 72
B-9000, Ghent, Belgium

Dr. D. Dixon-Holland	Neogen Corporation 620 Lesher Place Lansing, Michigan 48912 U. S. A.
Dr. A. C. E. Fesser	Health of Animals Laboratory 116 Veterinary Road Saskatoon, Saskatchewan S7N 2R3 Canada
Prof. J. P. Foley	Department of Chemistry Lousiana State University Baton Rouge Lousiana 70803 U. S. A.
Dr. M. Fujita	Department of Pharmaceutical Sciences Institute of Public Health 6-1, Shirokanedai 4 Chome, Minato-Ku, Tokyo Japan
Dr. A. De Guesquiere	Laboratory of Food Analysis University of Ghent Harelbekestrat 72 B-9000, Ghent Belgium
Dr. N. Haagsma	Department of Science of Food of Animal Origin Faculty of Veterinary Medicine University of Utrecht P. O. Box 80.175 3508 TD Utrecht The Netherlands
Dr. J. N. Harris	Medical Technology Corporation 71, Veronica Avenue Somerset, NJ 08873 U. S. A.
Dr. D. N. Heller	Division of Veterinary Medical Research Center for Veterinary Medicine Food and Drug Administration Barc East, Bldg. 328-A Beltsville, MD 20705 U. S. A.
Dr. M. Horie	Saitama Prefectural Institute of Public Health Saitama Japan
Dr. D. C. Holland	Animal Drug Research Center Denver Federal Building Food and Drug Administration Denver, CO 80225-0087 U. S. A.

Dr. J. A. Hurlbut	Animal Drug Research Center Denver Federal Building Food and Drug Administration Denver, CO 80225-0087 U. S. A.
Dr. T. Ida	The Public Health Laboratory of Chiba Prefecture 666-2, Nitona-Cho Chiba City Japan
Prof. S. E. Katz	Department of Biochemistry and Microbiology Cook College/New Jersey Agricultural Experiment Station Rutger-the State University of New Jersey New Brunswick, N.J. 08903-0231 U. S. A.
Dr. P. J. Kizak	Division of Veterinary Medical Research Center for Veterinary Medicine Food and Drug Administration Barc East, Bldg. 328-A Beltsville, MD 20705 U. S. A.
Dr. G. Korsrud	Health of Animal Laboratory 116 Veterinary Road Saskatoon, Saskatchewan, S7N 2R3 Canada
Dr. A. R. Long	Animal Drug Research Center Denver Federal Building Food and Drug Administration Denver, CO 80225-0087 U. S. A.
Dr. R. Malisch	Chem. Landesuntersuchungsantalt Freiburg Bissierstrabe 5 7800 Freiburg Fernsprecher 0761/8855-0 Germany
Dr. J. D. MacNeil	Health of Animal Laboratory 116 Veterinary Road Saskatoon, Saskatchewan, S7N 2R3 Canada
Dr. W. A. Moats	Meat Science Research Laboratory Agricultural Research Services USDA Beltsville, MD 20705 U. S. A.

Dr. R. K. Munns	Animal Drug Research Center Denver Federal Building Food and Drug Administration Denver, CO 80225-0087 U. S. A.
Ms. T. Nagata	The Public Health Laboratory of Chiba Prefecture 666-2, Nitona-Cho Chiba City Japan
Dr. H. Nakazawa	Department of Pharmaceutical Sciences Institute of Public Health 6-1, Shirokanedai 4 Chome, Minato-Ku Tokyo Japan
Dr. L. D. Payne	Department of Chemistry Merck, Sharp, and Dohme Reserach Laboratories P.O. Box 2000 (RY806L-123) Rahway, N.J. 07065
Prof. C. V. Peteghem	Laboratory of Food analysis University of Ghent Harelbekestrat 72 B-9000, Ghent Belgium
Dr. M. Petz	University of Wuppertal Department of Food Chemistry- FB9 Gau stra e 20 D-5600 Wuppertal-1 Germany
Dr. J. E. Roybal	Animal Drug Research Center Denver Federal Building Food and Drug Administration Denver, CO 80225-0087 U. S. A.
Dr. C. D. C. Salisbury	Health of Animal Laboratory 116 Veterinary Road Saskatoon, Saskatchewan, S7N 2R3 Canada
Dr. M. Saeki	The Public Health Laboratory of Chiba Prefecture 666-2, Nitona-Cho Chiba City Japan
Dr. S. Steiner	Medical Technology Corporation 71, Veronica Avenue Somerset, NJ 08873 U. S. A.

Dr. M. H. Thomas

Division of Veterinary Medical Research
Center for Veterinary Medicine
Food and Drug Administration
Barc East, Bldg. 328-A
Beltsville, MD 20705
U. S. A.

Dr. M. Waki

The Public Health Laboratory of Chiba Prefecture
666-2, Nitona-Cho
Chiba City
Japan

Dr. C. Van de Water

Department of Science of Food of Animal Origin
Faculty of Veterinary Medicine
University of Utrecht
P. O. Box 80.175
3508 TD Utrecht
The Netherlands

INDEX

Accuracy, 2, 3
Affinity, 24
Agarose, 85
Aglycone, 113
ß-agonists, 89
Agricultural, 23
Aklomide, 226
Albendazole, 123
Albumen, 24
Aldrin, 124
Alkaline phophatase, 59
Alkaloids, 212
Alkyl-α-D-penicilloic acids, 151
Alkyl sulfonate, 141
Alkylsulfonic acids, 155
Amino acids, 188
Aminoglycosides, 35, 48, 192
Ammonium sulphate, 25
Amoxicillin, 33, 60, 76, 133, 147
Amphoteric penicillin, 151
Amphoteric ß-lactam, 134
Ampicillin, 33, 60, 76, 124, 133,
        147, 189, 192
Amprolium, 190
Anabolic steroids, 99
Anabolic agents, 99
Anagyrine, 212
Antihelmintics, 226
Antibacterial, 71, 87, 188, 191
Antibiotic, 5, 157, 167
Antibody, 23
Antidote, 197
Antigen, 23, 57, 86
Antiglobulin reagent, 25
Antimethemoglobinemic, 197
Antimicrobial, 5, 6, 75
Anti-oxidants, 188
Antiparasitic, 113, 225
Antiseptic, 188, 197
Antisera, 25, 58
Apramycin, 46, 189
Avermectins, 62
Avoparcin, 189
Azure A, 197
Azure B, 197
Azure C, 197

B-o-galactosidase, 59
B. lymphosite, 82
B. stearothermophilus disc
        assay, 31
Bacillus brevis ATCC 8185, 192
Bacillus cereus ATCC 19637, 192
Bacillus cereus var. mycoides ATCC
        11778, 192
Bacillus stearothermophilus
        varcalidolactis C-953, 192
Bacillus stearothermophilus, 26
Bacillus subtilis ATCC 6633, 192
Bacitracin, 6, 46, 189
Bacteria, 19
Bambermycin, 6
Benzimidazoles, 122, 226
Benzthiazuron, 238
Benzyl penicillonic acid, 26
Benzylpenicillin, 147
Bicozamycin, 189
Biliary hyperplasia, 213
Bioassay, 191
Bioautography, 134
3,7-Bis(dimethylamino)phenothiazin-
        5-ium chloride, 197
Boldenone, 103
Bordetella bronchiseptica ATCC
        4617, 192
Bovine gamma globulin, 58
Bovine serum albumin, 26, 58
BR test, 34
4-Bromomethyl-7-methoxycoumarin,
        135, 155
Calcium halofuginone
        polystyrenesulfonate, 190
Carazolol, 82
Carbadox, 190, 226
Carbodiimide, 58
Carbomycin, 46, 189
Carbonyldiimidazole, 85
Carcasses, 99
Carcinogen, 165
Carcinogenity, 165
CAST, 7, 47
CDI-activated trisacryl, 85
Cefotaxime, 51

Ceftiofur, 33, 60, 76
Cellulose, 85
Centrilobular, 213
Cephalexin, 51, 83
Cephalosporins, 124
Cephapirin, 33, 60, 124, 133
Cepharonium, 189
Cephazolin, 189
Cephradine, 51
Charcoal, 32
Charm cowside test, 34
Charm farm test, 34
Charm test, 26
Charm II receptor assay, 17, 75
Chemotherapeutics, 225
Chloramphenicol, 8, 23, 32, 57, 82,
    107, 122, 189, 225
Chlorates, 197
Chlorhexidine, 46
Chlormadinone acetate, 103
Chlorotestosterone, 103
4-Chlorotestosterone, 104
4-Chlorotestosterone acetate, 103
Chlorotetracycline, 6, 33, 60, 108,
    125, 189, 192
Chlorsulan, 122
Chlorsulfuron, 122
Chromogenic, 198
CITE antibody test, 39
CITE test, 34
Clenbuterol, 89
Clopidol, 190, 226, 242
Cloxacillin, 33, 60, 76, 133, 147,
    189, 192
CNBr-activated sepharose, 85
Coccidiostats, 226
Colistin, 189, 192
Colorimetric detection, 134
Concomitant, 5
Cortisol, 91
Corynebacterium xerosis NCTC
    9755, 192
Coumaphos, 124
Crossreactivity, 59
Crown ethers, 139
18-Crown-6-ether, 155
Crufomate, 124
Cyanogen bromide, 85
Cyclobond-I, 167
Cyclodextrin, 171
D-alanine carboxypeptidase, 35
Dansylhydrazine, 153
Decoquinate, 190
Delvotest P, 34
Demeclocycline, 108
Demethylated metabolites, 197
Deprotenization, 134
Desoxycarbadox, 226
Destomycin A, 189, 192
Dextran coated charcoal, 27
Diatomaceous, 111

Diazotization, 58
Dicloxacillin, 155, 189, 192
Dicycloxacillin, 147
Dieldrin, 124
Dienestrol, 89, 103
Diethylether, 100
Diethylstilbesterol, 82
Diethylstilbesterol dipropionate,
    103
Dihydro-avermectin B1a, 113
Dihydro-avermectin B1b, 113
Dihydroerythroidines, 211
Dihydrostreptomycin, 46, 75, 189
Dimethyldichlorosilane, 100
Dimethylsulfoxide, 71
Dimetridazole, 225, 242
Dinitolamide, 226
Dinitrobenzamide, 226
Disc assay, 34
Disinfectant, 197
Doxycycline, 108, 189
E Z screen, 34
Electrochemical, 197
Electron capture detection, 27
ELISA, 17, 57
Emulsifiers, 188
Endrin, 124
Enramycin, 189, 192
Enzyme linked immunosorbent assay,
    57, 81
Enzyme immunoassay, 82, 191
Ergot, 213
Erythrina Berteroana, 211
β-Erythroidine, 212
Erythromycin, 6, 7, 33, 75, 148,
    189, 192
Escherichia coli, 198
Escherichia coli NIHG, 192
Estradiol 17 β, 103
Estradiol benzoate, 103
Estradiol 17 β-cypionate, 103
Estradiol 17 β-esters, 105
Estradiol 17 β-fenylpropionate, 103
Estradiol valerate, 103
Ethopabate, 190, 226
European economic community, 99, 134
Extra-label, 17, 32
Famfur, 124
FBZ-OH, 123
FBZ-SO₃, 123
Febantel, 225
Fenbendazole, 123, 225
Fenthion, 124
Ferricyanides, 197
Flavophospholipol, 189, 192
Flubendazole, 247
Fosfomycin, 189
Fradiomycin, 189, 192
Furazolidine, 124, 226
Gas chromatographic, 3, 26, 58, 150,
    166

Gas chromatography, 26, 191
Gas liquid chromatography, 134
GC/Electron capture, 128
GC/Mass spectrometry, 3, 89, 101,
    161, 227
Gentamicin, 6, 33, 60, 189
Gentian violet, 199
Glucose oxidase, 59
Glutaraldehyde, 58
Glycine/NaCl, 89
Halogenated pesticides, 122
Hepatocytes, 213
Hepatoxic pyrolizidine alkaloid, 21
Heptachlor, 124
Heptafluorobutyric acid anhydride,
    100
Hetacilline, 51, 60
Hexestrol, 89, 103
High performance liquid
    chromatographic, 165
High performance liquid
    chromatography, 27, 75,
    166, 191
High performance TLC plates, 100
Histomonostats, 226
Homogenization, 25, 126
Hormones, 23
Horseradish peroxidase, 59
HPLC, 16, 32, 37, 85, 138, 170, 227
HPLC receptogram, 37
Hydrazine, 135
Hydrogen peroxide, 62
8-a-Hydroxy-mutilin, 46
Hygromycin B, 6, 47, 189, 192
Imidazole, 153, 225
Immunization, 24
Immunoaffinity, 155
Immunoaffinity cleanup, 85
Immunoaffinity column, 87
Immunoassay, 60, 120, 126, 165
Immunochemical, 57
Immunochemical methods, 81
Immunogen, 24, 58
Immunoglobulin-G, 86
Immunoglobulins, 86,
Immunological, 19, 23
Immunopathological, 119
Immunoradiometric assays, 25
Immunosorbent, 86
Inclusion complex, 171
Incurred sample, 1, 2
Indolizidine, 213
Insulin, 23
Intrmuscular, 43
Ion supression, 139
Ipronidazol, 226
Isoxazolyl penicillins, 151, 157
Ivermectin, 62, 113, 122
Ivermectin-1-conalbumin, 62
Josamycin, 189
Kanamycin, 189, 192

Kasugamycin, 189, 192
Keyhole limpet hemocyanin, 27, 58
Kitasamycin, 189, 192
β-lactam, 16, 31, 57, 75, 133, 189
β-lactam antibiotic, 133
LacTek, 34
Lasalocid, 6
Lasalosid, 189, 192
LC/Mass spectrometry, 3
LC-MS, 161
Leucogentian violet, 199
Leucomethylene, 198
Ligands, 24
Lincomycin, 6, 35, 189
Lindane, 124
Lipid solubilizing polymer, 120
Liquid chromatographic, 3, 58
Liquid chromatography, 121
Macarbomycin, 189, 192
Macrolide, 16, 35, 192
Malachite green, 242
Mass-spectrometry, 3, 58, 191
Mass-specific detection, 27
Matrix solid phase dispersion, 108,
    119, 147
Mebendazole, 123
Mecillinam, 189
Medroxyprogesterone acetate, 103
Megestrol acetate, 103
Melengestrol acetate, 103
Mercuric mercaptide, 153, 158
Mestranol, 103
Meta-chloramphenicol, 240
Methacycline, 108
Methandienone, 103
Methicillin, 147
Methylation, 158
Methylboldenone, 104
Methylene blue, 197
Methylene chloride, 127
Methyltestosterone, 104
17-a-Methyltestosterone, 103
Meticlorpindol, 226
Metronidazol, 226
Microbial receptor assay, 31
Microbiological, 58, 119, 158
Micrococcus flavus ATCC 10240, 192
Micrococcus luteus ATCC 9341, 192
Microorganisms, 165
Mikamycin, 189
Minerals, 188
Minimum inhibitory concentration, 19
Minocycline, 108
Monensin, 6, 46, 83, 189, 192
Monobasic penicillin, 151
Monochlor chloramphenicol, 240
Monoclonal, 24, 58, 84
Monogastric animals, 212
Monosaccharide, 113
Morantel citrate, 190
Mutagenicity, 198

Mycotoxins, 213
Myeloma cells, 58
N4-acetylsulfaguanidine, 239
N4-acetylsulfapyridine, 226
N-hydroxy succinimide, 58
N-methyltrimethylsilyl-
    trifluoroacetamide, 100
Na nicotinamide, 238
Nafcillin, 51, 60, 147, 189
Nalidixic acid, 190, 225
Neomycin, 6, 33, 60, 75, 122
Nicarbazin, 122, 124, 190, 225
Nitrates, 197
Nitrofuranes, 225
Nitrofurazones, 226
Nitromide, 226
Norethisterone acetate, 103
17-α-nortestosterone, 87
17-β-nortestosterone, 87
19-Norethisterone, 103
19-Nortestosterone, 103
Nosiheptide, 189
Novobiocin, 33, 148, 189, 192
Nystain, 6
Nystatin, 189
O-phthaldialdehyde, 153
Olaquindox, 190
Oleandomycin, 6, 46, 75, 189, 192
Organophosphate, 124
Ormethoprim, 51
Oxacillin, 51, 60, 147
Oxfendazole, 123, 225
Oxolinic acid, 51, 190, 226
Oxytetracycline, 6, 33, 60, 108,
    125, 189, 192
Particle beam LC/MS interface, 114
Pathogenic bacteria, 119
Penaldic acid, 153
Penicillamine, 153
Penicillin, 6, 23, 60, 147
Penicillinase, 142, 147
Penicillin-G, 26, 133, 75, 133, 189,
    192
Penicillin V, 141
Penicilloyl group, 26
Penilloaldehyde, 153
Penzyme, 34
Penzyme II, 34
Periodate, 58
Peroxidase-antiperoxidase system, 83
Pharmacologic, 119
Phenoxymethylpenicillin, 147
Phosphate buffer saline, 85
Phospholipids, 138
Photodiode array detection, 149, 191
Photoreactor, 155
Physicochemical, 3, 81, 157
Piperacillin, 51
Piricularia oryzae, 192
Piromidic acid, 190, 226
Polyacrylamide, 85

Polyclonal, 24, 58, 83
Polyethylene glycol, 25
Polymixin B, 33, 189
Porcine tissue, 26
Post column electrochemical
    oxidation, 153
Post column reaction, 153
P,p'-DDE, 124
P,p'-DDT, 124
P,p'-TDE, 124
Practicability, 3
Precision, 2, 3
Precolumn derivatization, 135
Procaine, 76
Progesterone, 103
Prophylactic, 165
Proteolytic enzyme, 101
Pseudomonas syringae X 205, 192
Pyrazon, 238
Pyrimethamine, 190, 226
Pyrolizidine, 213
Quebemycin, 189, 192
Quinolizidine, 213
Radio receptor assay, 26
Radioactivity, 25
Radioimmunoassay, 23, 57, 82
Radioisotopes, 57
Radiolabelled, 75
Radiolabelled antigens, 23
Radiolabelled ligands, 23
Receptor assay, 133
Ribosomes, 35
Ronidazole, 242
S. Jacobaea, 213
Salinomycin, 6, 33, 46, 75, 148, 189
Sedecamycin, 189
Senecio alkaloids, 213
Senecio spp., 213
Slaughterhouses, 57
Sodium heptane sulfonate, 141
Solid phase extraction, 81, 90, 115,
    134, 158, 170
Solmonella typhimurium strains, 198
SOS, 7, 47
Specificity, 2, 3
Spectinomycin, 33, 189
Spiramycin, 189, 192
Spleen cells, 58
Stanozolol, 103
Staphylococcus aureus, 18, 19
Staphylococcus epidermidis ATCC
    12228, 192
STOP, 7, 47, 78
Streptavidine-biotine, 83
Streptomyces avermectilis, 71
Streptomycin, 192
Styrene divinylbenzene, 141
Subtherapeutically, 57
Subtilisin A, 101
Sulfaacetamide, 51
Sulfabromomethazine, 11, 46

Sulfachloropyridazine, 11, 32, 46, 165, 190
Sulfadiazine, 32, 125, 165, 190
Sulfadimethoxine, 11, 32, 60, 76, 122, 165, 173, 190, 241
Sulfadimidine, 190
Sulfadoxine, 51, 76, 190
Sulfaethoxypyridazine, 11
Sulfaguanidine, 226, 238
Sulfamerazine, 32, 63, 125, 165, 238
Sulfamethazine, 11, 32, 57, 75, 83, 125, 165, 238
Sulfamethiazole, 32, 165
Sulfamethoxazole, 125, 190, 241
Sulfamethoxypyridazine, 190
Sulfamonomethoxine, 173, 190
Sulfamoyldapsone, 190
Sulfanilamide, 51, 125, 165, 226
Sulfanitran, 46, 226
Sulfapyridine, 12, 33, 226, 238
Sulfaquinoxaline, 11, 33, 165, 190, 241
Sulfathiazole, 11, 33, 60, 76, 125, 165, 190, 238
Sulfisoxazole, 51, 125, 165, 190
Sulfomyxin, 46
Sulfonamides, 6, 57, 75, 225
Supercritical fluid chromatography, 134
Tanzy ragwort, 213,
Testosterone, 89, 103
17 α-Testosterone, 104
17 β-Testosterone, 104
Testosterone cypionate, 103
Testosterone enantate, 103
Testosterone fenylpropionate, 103
Testosterone isocaproate, 103
Testosterone propionate, 103
Tetracycline, 6, 32, 33, 60, 108, 122, 189
Tetracycline-HCl, 46
Tetraethyl ammonium chloride, 141

Tetrahydroerthroidine, 215
Tetrahydrofuran, 71
Tetramethylbenzidine, 62
Tetrathionine chloride, 197
Therapeutic, 147, 165
Therapeutically, 57
Thermionic nitrogen selective detector, 157
Thermospray HPLC-MS, 195
Thermospray-MS, 114
Thiabendazole, 123
Thiamphenicol, 190
Thin layer chromatography, 191
Thin layer chromatographic, 3, 58, 134, 166
Thionin, 197
Thiopeptin, 189
Tiamulin, 46, 189
Ticarcillin, 51
Tolerance levels, 6
Toluene, 100
Toluidine blue, 197
Toxicologic, 119
Toxins, 212
Trenbolone, 89, 104
Trenbolone acetate, 103
Triazole, 135, 153
Triazole mercuric chloride, 135
Trimethoprim, 76, 190
Trimethylchlorosilane, 100
Trimethylsilyliodide, 100
Trisacryl, 85
Trishydroxymethylaminomethane, 100
Tungstic acid, 134
Tylosin, 6, 33, 148, 189, 192
Ultrafiltration, 134, 147
Veterinary, 23
Vinyltestosterone, 103
Virginiamycin, 6, 46, 189, 192
Vitamins, 188
Zeranol, 103
Zoalene, 226